Springer Tracts in Mechanical Engineering

Springer Tracts in Mechanical Engineering (STME) publishes the latest developments in Mechanical Engineering - quickly, informally and with high quality. The intent is to cover all the main branches of mechanical engineering, both theoretical and applied, including:

- Engineering Design
- Machinery and Machine Elements
- Mechanical Structures and Stress Analysis
- Automotive Engineering
- Engine Technology
- Aerospace Technology and Astronautics
- Nanotechnology and Microengineering
- Control, Robotics, Mechatronics
- MEMS
- Theoretical and Applied Mechanics
- Dynamical Systems, Control
- Fluids Mechanics
- Engineering Thermodynamics, Heat and Mass Transfer
- Manufacturing
- Precision Engineering, Instrumentation, Measurement
- Materials Engineering
- Tribology and Surface Technology

Within the scope of the series are monographs, professional books or graduate textbooks, edited volumes as well as outstanding PhD theses and books purposely devoted to support education in mechanical engineering at graduate and post-graduate levels.

Indexed by SCOPUS, zbMATH, SCImago.

Please check our Lecture Notes in Mechanical Engineering at https://link.springer.com/bookseries/11236 if you are interested in conference proceedings.

To submit a proposal or for further inquiries, please contact the Springer Editor **in your region**:

Ms. Ella Zhang (China)
Email: ella.zhang@springernature.com
Swati Meherishi (India)
Email: swati.meherishi@springer.com
Dr. Leontina Di Cecco (All other countries)
Email: leontina.dicecco@springer.com

All books published in the series are submitted for consideration in Web of Science.

Khrystyna Berladir · Zuzana Mitaľová ·
Vitalii Ivanov

Materials for the Automotive Industry

Springer

Khrystyna Berladir
Technical University of Košice
Prešov, Slovakia

Sumy State University
Sumy, Ukraine

Vitalii Ivanov
Technical University of Košice
Prešov, Slovakia

Sumy State University
Sumy, Ukraine

Zuzana Mitaľová
Technical University of Košice
Prešov, Slovakia

ISSN 2195-9862 ISSN 2195-9870 (electronic)
Springer Tracts in Mechanical Engineering
ISBN 978-3-032-05865-2 ISBN 978-3-032-05866-9 (eBook)
https://doi.org/10.1007/978-3-032-05866-9

This Springer imprint is published by the registered company Springer Nature Switzerland AG
The registered company address is: Gewerbestrasse 11, 6330 Cham, Switzerland

Preface

The automotive industry stands at the intersection of technological innovation and material science, driving progress and shaping the future of transportation. With the rapid advancements in vehicle design and manufacturing, the role of materials has never been more crucial. This book, organized into 12 comprehensive chapters, delves into the diverse materials that form the backbone of modern automotive engineering.

Chapter 1: Introduction provides a brief overview of the current state of research on the materials that define the automotive landscape. Also, the chapter examines the key materials used in the industry.

Chapter 2: Steel describes the different classifications of automotive steels, including advanced high-strength and stainless steels. Steel's role as a fundamental material in vehicle construction is underscored by its versatility and strength.

Chapter 3: Cast Iron presents the classification and covers the various types of cast irons, their mechanical properties, and their specific applications in automotive components. The chapter emphasizes graphite iron, ductile iron, and special cast iron used in the automotive industry.

Chapter 4: Aluminum and Its Alloys presents the historical evolution of aluminum from 1899 to the present day. It highlights aluminum's lightweight yet strong properties, making it a preferred material for car bodies and other critical parts.

Chapter 5: Magnesium and Its Alloys emphasizes the benefits of magnesium in reducing vehicle weight without compromising strength. The chapter delivers the list of manufacturers of magnesium alloy components and applications for automotive products.

Chapter 6: Copper and Its Alloys delves into copper's electrical and thermal conductivity advantages, making it indispensable in various automotive systems. Numerical and graphical data provide a better understanding of the rational use of this type of material.

Chapter 7: Titanium and Its Alloys brings to light the high-strength, lightweight properties of titanium, which are increasingly used in high-performance and luxury vehicles. It is worth admitting that the development of titanium started with F-1 racing cars and moved to serial automotive vehicles.

Chapter 8: Plastics discusses the evolution, types, and sustainable alternatives of plastics in the automotive industry, reflecting the growing importance of environmental considerations. Different types of plastics (e.g., polypropylene, polyamide, acrylonitrile-butadiene-styrene, etc.) are widely used in the design of automotive components, mainly for internal design.

Chapter 9: Biocomposites signals a move towards greener materials, highlighting natural fibers and their applications in automotive design. The chapter introduces the classification and raw materials of biocomposites with natural fibers, highlighting their properties and the rational use area.

Chapter 10: Glass examines the different types of automotive glass and the complexities of glazing systems essential for vehicle safety and aesthetics. Nowadays, glass has become a functional product that needs advanced manufacturing processes and technological equipment.

Chapter 11: Rubber and Rubber Materials explores the critical role of rubber in the automotive industry, covering various rubber materials, their unique properties, and wide-ranging applications. From tires to seals and gaskets, rubber's versatility and resilience make it indispensable in ensuring vehicle safety, performance, and durability. The chapter also delves into the advancements in rubber technology that continue to enhance automotive design and functionality.

Chapter 12: Smart Materials offers a glimpse into the future with materials like piezoelectric substances and shape memory alloys that respond to environmental changes, paving the way for the next generation of vehicles.

This book is designed to serve as a comprehensive reference for students, researchers, and professionals in the automotive industry, providing a detailed understanding of the materials that drive innovation. As the automotive industry continues to evolve, so must our knowledge of the materials that make it possible. This book aims to equip readers with the necessary insights to navigate and contribute to this dynamic field.

The entire material presented in the textbook is based on the authors' scientific experience at the Technical University of Košice (Prešov, Slovak Republic) and Sumy State University (Sumy, Ukraine).

Preparation and publication of this book was realized within the research, funded by the EU NextGenerationEU through the Recovery and Resilience Plan for Slovakia under the projects No. 09I03-03-V01-00101 and No. 09I03-03-V01-00094. The book preparation was partially supported by the International Association for Technological Development and Innovations and the International Innovation Foundation.

Prešov, Slovakia

Khrystyna Berladir
Zuzana Mitaľová
Vitalii Ivanov

Competing Interests The authors have no competing interests to declare that are relevant to the content of this manuscript.

Contents

About the Authors

Dr. Khrystyna Berladir Ph.D., Associate Professor. She works at the Department of Automobile and Manufacturing Technologies of the Technical University of Košice (Slovak Republic) and the Department of Applied Materials Science and Technology of Structural Materials of Sumy State University (Ukraine). Her research activities concern Materials Science issues, including the design, simulation, and technology of composite materials, materials modification for the automotive industry, biomaterials and bioengineering, quality management for material selection in product development, and innovations in engineering education. Dr. Berladir completed internships at Poznan University of Technology (Poland) and the University of Liverpool (UK). She also gave lectures within the Erasmus program at Eskişehir Technical University (Turkey), University of West Bohemia (Czech Republic), Lublin University of Technology (Poland), Alexander Dubček University of Trenčin (Slovak Republic). She is a member of the International Association for Technological Development and Innovations (IATDI), the International Innovation Foundation (IIF), and the European Alliance for Innovation (EAI). Author of over 100 publications in materials science, Dr. Berladir is also acting as a reviewer of scientific articles published in Springer Nature, Elsevier, and MDPI publishing houses. She is also a member of the program and organizing committees of international conferences Design, Simulation, Manufacturing: The Innovation Exchange (DSMIE, 2018–2025), Grabchenko's International Conference on Advanced Manufacturing Processes (InterPartner, 2019–2025), scientific committee of the international conference MANUFACTURING (2022, 2024). Dr. Berladir is the co-editor of the proceedings of the 2nd and 3rd EAI International Conferences on Automation and Control in Theory and Practice (2024 and 2025).

Dr. Zuzana Mitaľová Ph.D., Associate Professor. She completed B.Sc. and M.Sc. studies at the Faculty of Manufacturing Technologies of Technical University of Košice with a seat in Prešov. In 2012, she defended a Ph.D. thesis titled: "A Contribution to Solving the Problem of Machining Composite Materials Based on WPC". In 2014, she completed her postdoctoral studies at VŠB-TU in Ostrava, Czech

Republic. In 2023, she was appointed as an associate professor in the field of Manufacturing Technologies. She currently works at the Faculty of Manufacturing Technologies with a seat in Prešov at Technical University of Košice. She is the author of more than 100 scientific publications published in both foreign and domestic publishers and co-author of two patents. Her scientific and professional focus is oriented toward composite materials with natural reinforcements, biomaterials, and material applications in the automotive industry.

Prof. Vitalii Ivanov D.Sc., Ph.D., ING.PAED.IGIP, Head of the Department of Manufacturing Engineering, Machines and Tools at Sumy State University, and Research Professor at the Department of Automobile and Manufacturing Technologies of the Technical University of Košice (Slovak Republic). He was awarded a Scholarship of the Cabinet of Ministers of Ukraine for Young Scientists (2011 and 2014), a Fellowship for a Research Internship from the Ministry of Education and Science, Youth and Sports of Ukraine (2011 and 2012), The National Scholarship of the Slovak Republic (2015, 2017, and 2022), JICA Knowledge Co-Creation Fellowship (2016), The Ulam NAWA Programme (2022), and State Prize of the Supreme Council of Ukraine for Young Scientists (2022). Prof. Ivanov carried out research, completed internships, and gave lectures in Germany, USA, Slovak Republic, Poland, Great Britain, Italy, Japan, Czech Republic, Romania, China, etc. Furthermore, he participated in research projects funded by the Ministry of Education and Science of Ukraine, State Fund for Fundamental Research, and Ukrainian and German industrial companies. He also has experience coordinating and co-coordinating research projects funded by TEMPUS, Erasmus+, Czech Development Agency, DAAD, Gene Haas Foundation and Haas Automation, Japan International Cooperation Agency, and NAWA. Since 2019, Prof. Vitalii Ivanov has been a Member of the Advisory Board for Research Evaluation at the University of West Bohemia (Czech Republic). Since 2020, he has been a Member of the Experts' Team of the Invention Support Fund of the Ministry of Economic Development and Trade of Ukraine and an Expert of the Ministry of Education and Science of Ukraine on the State Certification of Scientific Institutions. In 2022, he became a Member of the Expert Council of the Ministry of Education and Science of Ukraine on certification of scientific personnel in Manufacturing Engineering. In 2021, Prof. Ivanov became an International Member of the Academic Committee of the Doctoral Program "Innovative Engineering Technologies for Industrial Sustainability—IETIS" at the University of Salerno (Italy). In 2022, Prof. Ivanov became a Member of the Specialized Councils for obtaining a Doctor of Science Degree in 05.02.08—Machine Building Technology, 05.03.01—Machining Processes, Machines, and Tools at National Technical University "Kharkiv Polytechnic Institute" (Ukraine) and Odesa Polytechnic National University (Ukraine). Since 2024, he has become an Expert in carrying out scientific evaluation in Manufacturing and Mechanical Engineering (Ministry of Education and Science of Ukraine).

Chapter 1
Introduction

Abstract This chapter presents an overview of the materials used in the automotive industry, presenting their evolution and current trends. It highlights transitioning from traditional steel-based construction to a mixed-material approach, followed by performance, safety, fuel efficiency, and regulatory pressures. Key materials include various grades of steel, aluminum, magnesium, titanium, composites, and smart materials. The shift toward lightweight, high-strength materials aims to reduce vehicle weight by up to 75%. However, cost, manufacturing complexity, and joining techniques remain critical challenges. Steel maintains dominance due to its strength and cost-effectiveness despite the rise of advanced materials. The chapter concludes that future automotive design will integrate diverse materials optimized for specific applications, supported by new manufacturing technologies and joining methods to meet evolving industry demands.

1.1 Current State of the Problem

The automotive industry is well-developed in the U.S., Western Europe, Brazil, Australia, and East Asia [1]. In 1960, the U.S. produced more than 50% of the world's cars, and Great Britain—10%. Their share decreased to 18 and 3.3% three decades later, respectively. Japan overtook these countries. In 1960, it produced 165,000 cars; in 1991, it produced thirteen million, and its share now accounts for 25% of the vehicles produced worldwide [2]. 40–50 million cars are produced annually globally, four of which are trucks and buses.

This sphere is the leader in the number of multinational companies. About ten leading companies produce 80% of the world's automobile production [3]. For example, the American company «Ford» is the second largest car manufacturer in the world; 60% of its cars are produced outside the U.S.

More than half of the total volume in the production of a modern passenger car consists of parts made of cast iron and steel (55%), the second place—plastics (about 11%), the third place—aluminum alloys (9%); rubber and glass—7 and 3%, respectively [4]. The share of non-ferrous alloys (magnesium, titanium, copper, and

K. Berladir et al., *Materials for the Automotive Industry*, Springer Tracts in Mechanical Engineering, https://doi.org/10.1007/978-3-032-05866-9_1

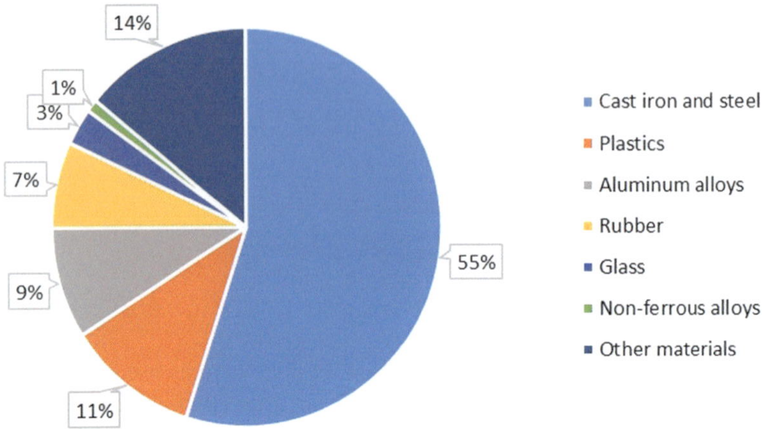

Fig. 1.1 Classes of materials used to produce cars

zinc) does not exceed 1%; other materials (varnishes, paints, electrical wires, facing materials, etc.) make up 13.5% (Fig. 1.1).

Developing new materials with improved performance is driven by growing demands in the automotive industry to improve impact resistance, noise, and vibration and reduce overall cost and fuel economy.

Design engineers try to implement the concept of «the right material in the right place» [5]. Theoretically, any material can be used for the needs of automobile construction if it meets the following requirements:

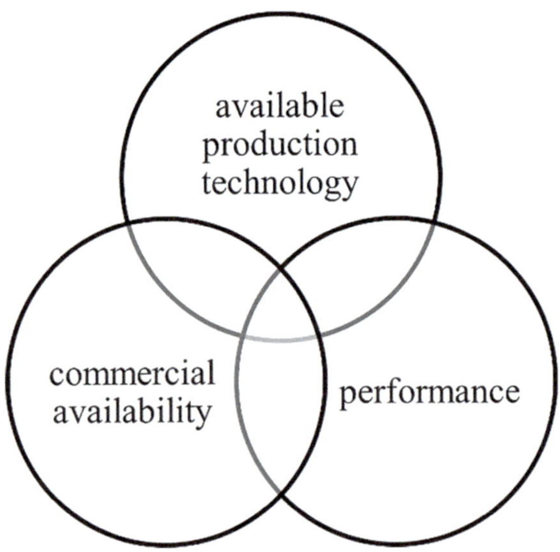

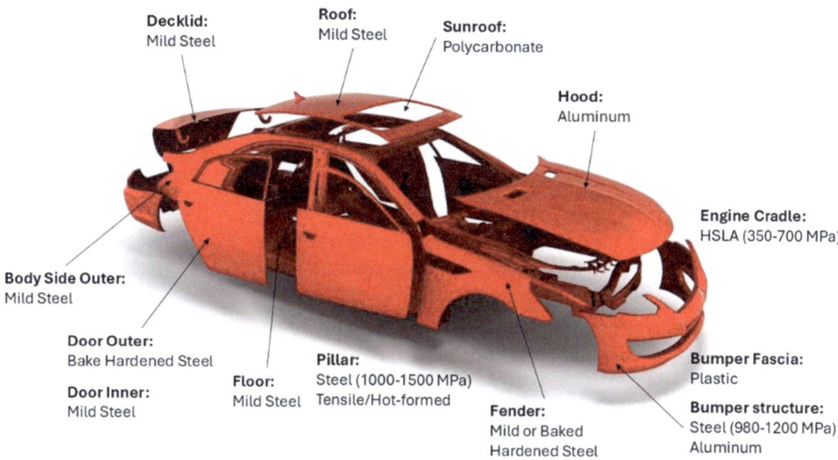

Fig. 1.2 Materials for major structural components of a typical automobile construction

However, designers may not use all available materials due to day-to-day constraints (supply chain, infrastructure, cost, repairability, environment, etc.) [6]. Materials that are currently most used for major structural components of typical automobile construction are shown in Fig. 1.2.

Currently, and predicted to be soon, the most used automotive materials are described in Table 1.1.

Automotive structures mainly have metal parts made of aluminum. Steel grades range from mild to hot-formed boron. Composites based on magnesium and polymers are used in some components, primarily high-end cars. The current combination of materials in the main components of the car, from the body structure to the closing means [11], is shown in Fig. 1.3.

The race to improve performance, as well as regulatory pressure to reduce carbon dioxide emissions, drive the shift in the automotive industry's material mix. Automakers are interested in materials with a higher strength-to-weight ratio, which reduces weight and improves vehicle performance.

For example, the tensile properties of a few selected materials already used in current autos compete with steel and are likely to be used for future vehicle construction are presented in Table 1.2.

Many of them are more expensive than steel; there are technical and cost challenges to be overcome for large production volumes. Despite this, many concept vehicles and prototypes have demonstrated their technical viability and weight-saving potential. Even, they appear in some production vehicles, but in smaller quantities.

CAR research predicted that by 2025, the U.S. vehicle fleet will be reduced by five percent due to the increased use of aluminum. The composition of automotive products is rapidly changing and becoming incredibly diverse [13]. From 2020 to 2030, mild steel will be halved and replaced by stronger steel, aluminum, and composites

Table 1.1 The most used classes of automotive materials [7–10]

Material	Type of material	Characteristics	Typical application
Steels	Mild steel	It is easily formed using cold forging and other outdated manufacturing processes. The maximum tensile strength is 270 MPa	Frame, body, crash zone, pillar, roof rail, door beam, seat frame, front side member, roof bow, bumper reinforcement, rocker, cross member, seat track
	High-strength steel (HSS)	HSSs are traditional soft steels strengthened by thermal treatment. The tensile strength ranges from 250 to 550 MPa	
	High strength low alloy (HSLA)	HLSAs are carbon-manganese steels strengthened by adding a microalloying element such as Ti, V, or Nb. The maximum tensile strength is 800 MPa	
	Advanced high-strength steel (AHSS)	AHSSs are composites of multiple metals heated and cooled throughout the manufacturing process to meet a part's specifications. The yield strength level is more than 550 MPa	
	Ultra-high strength steel (UHSS)	UHSSs have similar properties as AHSSs. However, the minimum tensile strength is 780 MPa	
	Boron/Martensite	Martensite is the hardest and strongest form of steel. It's also the least malleable. It shares properties with boron, with a tensile strength of about 1200–1800 MPa. They are usually combined with mild steel to form composites	
Light alloys	Aluminum 5000/6000 (AL 5000/6000)	AL 5000 contains Mg. AL 6000 contains both Si and Mg, which forms magnesium silicide and makes the aluminum alloy heat-treatable	Shock absorber, piston, brake, tank, fender, roof, door, suspension component, bumper, chassis, heat insulator, handle, piping, wheel rim, steering component, conrod, rotor, bonnet, spoke, valve, gas cylinder, seat frame

(continued)

Table 1.1 (continued)

Material	Type of material	Characteristics	Typical application
	Magnesium (Mg)	Mg is a lightweight material often used for automotive. In alloys, magnesium has the highest strength-to-weight ratio of all structural metals	The engine block, seat frame, steering wheel frame, wheel rim, instrument panel, clutch case, cylinder block, cylinder head, transmission case, lower crankcase, air intake system, camshaft drive chain case, steering link bracing, intake manifold, oil pump body, gear control housing, bracket
	Titanium (Ti)	Ti alloy has excellent corrosion resistance compared to stainless steel, especially in marine atmospheric environments, chloride ion erosion, and micro-oxidation atmospheres. It also maintains good ductility, and the service temperature of refractory titanium alloy can reach around 550 °C, making it significantly more heat resistant than aluminum and magnesium alloy	Connecting rod, spring, engine valve, wheel, intake valve, exhaust system, turbocharger, body frame, muffler, engine rocker arm, engine piston pin, suspension component, lug nut, fastener, car stop bracket, door penetration beam, pin bolt, pressure plate, brake caliper piston, shift button, fuel tank, clutch circle, fuel cell separator
Composites	Carbon fiber reinforced plastic (CFRP)	CFRP contains carbon fibers to increase strength. They are extremely strong and light plastics but expensive to produce	Bumper beam, frontal bonnet, body, door panel, seatback, cargo area, lining, instrumental panel, dashboard, insulation, body panel, wheel box, pillar, windshield, floor tray, roof cover, hood, and rear storage shelf
	Glass fiber reinforced plastic (GFRP)	GFRP is a composite material with fiber, E-glass, and polyester. The tensile strength ranges from 44 to 2358 MPa, and a compressive strength from140 to 350 MPa; while only weighing a quarter of the weight of steel	
Smart materials	Piezoelectric ceramics, shape memory alloys, magnetostrictive materials, nanomaterials, etc.	Smart materials can change shape, positioning, flexibility, fluid flow rate, etc. They can respond quickly to temperature, electric field, and magnetic field changes	Various applications are based on material-specific properties

Fig. 1.3 The used material mix for major components from mass-produced cars

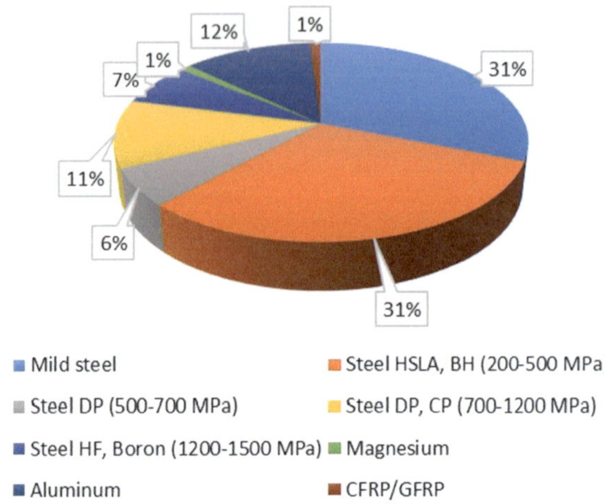

(Fig. 1.4). This trend will continue to evolve as materials become more advanced and cost-competitive with traditional steel.

It is believed that reducing the weight of an average car by 10% leads to a decrease in fuel consumption by 6–8%. At the same time, the body (40%), chassis (25%), equipment (20%), and power plant (15%) make up the average weight of the car. The car's weight is reduced by replacing traditional metal alloys with modern, stronger, lighter materials (Table 1.3).

Conventional high-strength, low-alloy steels (HSS) retain economic advantages over other lightweight materials. At the same time, composite materials remain expensive due to their high cost and long production cycle. The slow growth in the use of aluminum instead of steel is due to a significant difference in their cost. A cost–benefit analysis shows that steel parts are stronger and cheaper than equivalent parts made from other lightweight materials.

No single material is predicted to win the relief race. Optimized for weight and performance, the car will feature a mixed-material body structure. The industry is already experiencing this change with newly introduced vehicles that use materials tailored to each part of the vehicle to improve driving dynamics, fuel economy, and cabin quietness.

New manufacturing technologies are also being developed to achieve the speed and cost-effectiveness required for mass production [11]. For example, hot forming of steel is already used in high-performance parts and will reach maturity by 2025 as the need for ultra-high-strength steel increases. Cold-formed steel is still the industry standard for automotive parts production, but as detailed above, high-strength materials are the future. High-strength steels are difficult to cold form, leading to the recent development of hot forming or hot stamping, where stronger, thinner, and lighter components can be developed.

Innovative production processes that will rethink the production of automotive parts are presented in Fig. 1.5.

Table 1.2 Material property comparisons [12]

Material	Density (ρ), (g/cm^3)	Tensile modulus (E), (GPa)	Yield strength (S_y), (MPa)	Tensile strength (S_t), (MPa)	Coefficient of thermal expansion (10^{-6}/°C)
DQ low carbon steel	7.87	207	186	317	11
DP 400/700 steel	7.87	207	400	700	11
TRIP 450/800 steel	7.87	207	405	800	11
5182-H24 aluminum	2.7	70	235	310	23
6111-T62 aluminum	2.7	70	320	360	23
AZ91 magnesium	1.8	45	160	240	26
Ti-6Al-4 V titanium	4.43	114	827	896	9
304 stainless steel	7.9	200	241	614	17
Nitronic 30 stainless steel	7.86	193	393	862	16
High-strength CFRE (unidirectional)	1.55	138	–	1550	–
High modulus CFRE (unidirectional)	1.63	215	–	1240	–0.9 (L); 27 (T)
GFRE (unidirectional)	1.85	39	–	965	6 (L); 19 (T)
CFRE (quasi-isotropic)	1.55	45.5	–	579	0.9
Sheet molding compound (SMC-R50)	1.87	16	-	164	14.8

Note: L is the longitudinal direction, and T is the transverse direction
(1) CFRE is carbon fiber-reinforced epoxy, and (2) GFRE is glass fiber-reinforced epoxy

Production of new materials requires new challenges. Joining dissimilar materials is not easy and sometimes impossible with traditional contact spot welding due to differences in melting points. Joining technologies such as adhesives and advanced fasteners will play an important role in achieving optimized mixed-material architecture, as they can join any combination of dissimilar materials.

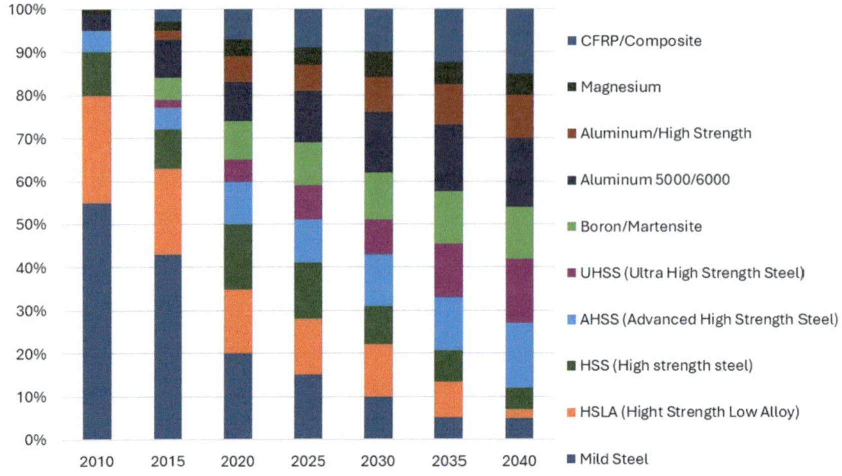

Fig. 1.4 Forecast of the distribution of materials used in the U.S. fleet

Table 1.3 Weight and cost indicators of replacing traditional metal alloys with new materials [14]

Replaceable traditional alloys	New strong and light materials	Details of strong and light materials	
		Weight reduction, %	Relative cost, unit
Low carbon steel	HSS	10–25	1
Aluminum	Magnesium	25–35	1–1.5
Steel	Polymer composite materials reinforced with glass fiber	25–35	1–1.5
Carbon steel	Stainless steel	20–45	1.2–1.7
Steel, cast iron	Aluminum	40–60	1.3–2.0
Foundry steel	Titanium	40–55	1.5–10+
Steel or cast iron	Magnesium	60–75	1.5–2.5
Steel or cast iron	Aluminum-matrix composites	50–65	1.5–3+
Steel	Polymer composite materials reinforced with carbon fiber	50–60	2–10+

1.2 Summary

The automotive industry is in a state of continuous evolution, driven by the need for improved performance, safety, and fuel efficiency. The transition from traditional steel-based vehicle structures to a diverse mix of materials, including high-strength steel, aluminum, magnesium, and composites, highlights the industry's pursuit of weight reduction and enhanced durability. As regulatory pressures and consumer

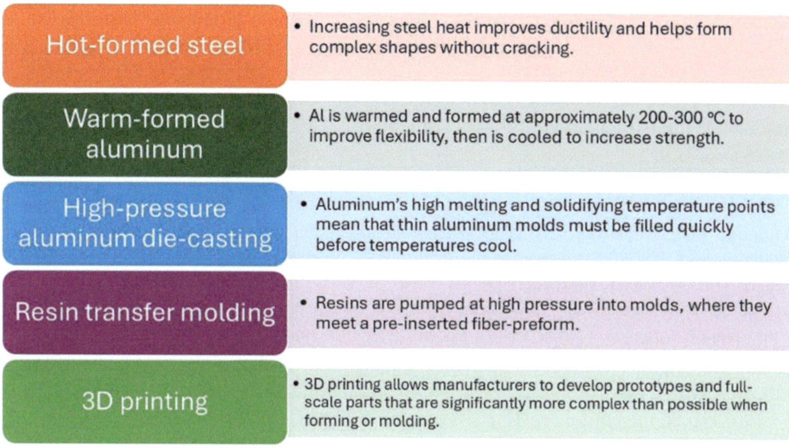

Fig. 1.5 Future production of automotive parts

expectations grow, automakers are focusing on materials that offer a better strength-to-weight ratio while maintaining cost-effectiveness and manufacturability.

Despite the increasing use of lightweight materials, steel remains a dominant choice due to its affordability, strength, and established manufacturing processes. However, advancements in alternative materials such as carbon fiber composites and titanium alloys are proving their potential in specialized applications, particularly in high-performance and luxury vehicles. The challenge remains in balancing cost, production feasibility, and the integration of these advanced materials into mass-market vehicles while ensuring safety and structural integrity.

Looking ahead, the automotive industry is expected to embrace a mixed-material approach, utilizing each material's unique advantages for specific vehicle components. This shift will be accompanied by innovations in joining techniques, manufacturing processes, and supply chain strategies to accommodate the changing material landscape. As technology progresses, industry must continue to find sustainable and economically viable solutions to meet future demands while enhancing vehicle performance and environmental sustainability.

References

1. Graham J (2010) The global economic crisis: sectoral coverage: automotive industry: trends and reflections. International Labour Organization. https://researchrepository.ilo.org/esploro/outputs/encyclopediaEntry/The-global-economic-crisis-sectoral-coverage/995218816902676. Accessed 21 Jan 2025
2. Humphrey J, Memedovic O (2003) The global automotive industry value chain: what prospects for upgrading by developing countries. UNIDO Sectorial Studies Series Working Paper. SSRN. https://ssrn.com/abstract=424560. Accessed 21 Jan 2025

3. Anagor-Ewuzie A (2024) 10 world's biggest automobile-producing countries in 2023. https://businessday.ng/transport/article/10-worlds-biggest-automobile-producing-countries-in-2023/. Accessed 21 Jan 2025
4. Hovorun T P, Berladir K V, Pererva V I, Rudenko S G, Martynov A I (2017) Modern materials for automotive industry. Journal of Engineering Sciences 4(2):F8–F18. https://doi.org/10.21272/jes.2017.4(2).f8
5. Ashby MF (2004) Materials selection in mechanical design. Butterworth-Heinemann, UK. https://doi.org/10.1016/C2009-0-25539-5
6. Maleque MA, Salit MS (2013) Overview of materials selection and design. In: Materials selection and design. Springer Briefs in Materials. Springer, Singapore, pp 1–15. https://doi.org/10.1007/978-981-4560-38-2_1
7. Zhang W, Xu J (2022) Advanced lightweight materials for automobiles: a review. Mater Des 221:110994. https://doi.org/10.1016/j.matdes.2022.110994
8. Khan F, Hossain N, Mim JJ, Rahman SMM, Iqbal MdJ, Mostakim Billah M, Chowdhury MA (2024) Advances of composite materials in automobile applications – a review. J Eng Res 13:1001–1023. https://doi.org/10.1016/j.jer.2024.02.017. (in press)
9. Salonitis K, Pandremenos J, Paralikas J, Chryssolouris G (2009) Multifunctional materials used in automotive industry: a critical review. In: Pantelakis S, Rodopoulos C (eds) Engineering against fracture. Springer, Dordrecht, pp 59–70. https://doi.org/10.1007/978-1-4020-9402-6_5
10. Czerwinski F (2021) Current trends in automotive lightweighting strategies and materials. Materials 14:6631. https://doi.org/10.3390/ma14216631
11. Smith B, Spulber A, Modi S, Fiorelli T (2017) Technology roadmaps: intelligent mobility technology, materials and manufacturing processes, and light duty vehicle propulsion. Center for Automotive Research. https://www.cargroup.org/publication/technology-roadmaps-intelligent-mobility-technology-materials-and-manufacturing-processes-and-light-duty-vehicle-propulsion/. Accessed 21 Jan 2025
12. Mallick PK (2012) 2 – Advanced materials for automotive applications: an overview. In: Rowe J (ed) Advanced materials in automotive engineering. Woodhead Publishing, UK, p 5–27. https://doi.org/10.1533/9780857095466.5
13. Next-Generation Automotive Manufacturing Materials and Processes (2022). https://funding.ryan.com/blog/business-strategy/automotive-manufacturing-materials-processes/?utm_source=mentorworks.ca&utm_medium=referral&utm_campaign=redirect. Accessed 21 Jan 2025
14. Gnatush VA, Doroshenko VS (2019) Modern use of metal alloys for the automotive industry. Metal Cast Ukr 27(10–12):74–81. https://doi.org/10.15407/steelcast2019.10.074

Chapter 2
Steel

Abstract This chapter describes steel's properties, classifications, and applications in the automotive industry, emphasizing its central role due to its strength, cost-effectiveness, and adaptability. It outlines the evolution from conventional steels to advanced high-strength steels, including dual-phase, transformation-induced plasticity, complex phase, martensitic, and twinning-induced plasticity steels. Special attention is given to structural performance, formability, and crash resistance. The chapter also details specific chemical compositions, microstructures, and mechanical properties of progressive steel types and highlights ongoing innovations. Steel remains the dominant material for automotive structures due to continuous advancements in its performance and manufacturing capabilities.

2.1 Introduction

Steel is an alloy of iron and carbon with improved strength and fracture resistance compared to other forms of iron [1]. Many other elements may be present or added. Stainless steels that are corrosion and oxidation-resistant typically need an additional 11% chromium. Because of its high tensile strength and low cost, steel is used in buildings, infrastructure, tools, ships, trains, cars, bicycles, machines, electrical appliances, and weapons [2].

Steel is used extensively in cars and trucks. The metal is used for various purposes, from the frame to the body to the undercarriage. Steel is strong and durable, making it ideal for vehicle construction. It can also be worked into various shapes, allowing manufacturers to create complex designs.

Steel is used in various ways in the automotive industry as a versatile material [3]. It is often used in the chassis and body of vehicles, as well as in the manufacturing of engine parts. Steel is strong and durable, making it an ideal choice for many automotive applications. Rolled steel plays a leading role in production despite the increased use of new structural materials in the automotive industry. The automotive industry places a high demand for steel because it must first meet two opposing criteria. On the one hand, the requirement to reduce the weight of products implies

K. Berladir et al., *Materials for the Automotive Industry*, Springer Tracts in Mechanical Engineering, https://doi.org/10.1007/978-3-032-05866-9_2

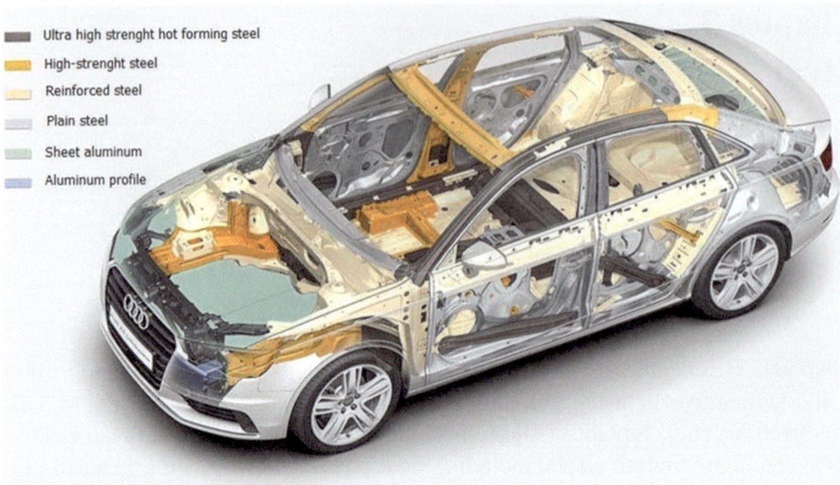

Fig. 2.1 Materials used to manufacture car bodies [4]

the use of high-strength materials; on the other hand, the increase in requirements for production manufacturability implies the use of highly plastic materials [4].

Steel is a time-tested material. This material has been used in the manufacture of bodies for a long time due to its strength (Fig. 2.1).

In 1994, 35 steel-producing companies from 18 countries developed and began to implement the UltraLight Steel Autobody (ULSAB) program to optimize the environmental and weight parameters of cars and in opposition to manufacturers of aluminum alloys [5]. WorldAutoSteel, the automotive group of the World Steel Association (WSA), actively participated in implementing the ULSAB program. The successful implementation of the specified program served as the basis for further progress and the initiation of two more programs [6] (Table 2.1).

According to the ULSAB program, the car body consists of 94 main parts, excluding the doors and hood, and weighs 203 kg, 25% less than the average base body. Such positive results were achieved due to the development of a line of deformed automotive steels, which are divided into four groups depending on the strength limit (Table 2.2).

Thus, the use of steel materials for cars has its advantages and disadvantages (Fig. 2.2).

2.2 Classification of Automotive Steels

Automotive steels can be classified in several different ways [2].

Table 2.1 Programs for global automotive and steel manufacturers

Program	Period of validity	Indexes
UltraLight steel Autobody, ULSAB	1994–1998	35 steel-producing companies in the world took part in the consortium Reduction of car weight by 25%, development of high-strength (high-strength steels, HSS) and ultrahigh-strength (ultrahigh strength steels, UHSS) steel grades
UltraLight Steel Autobody-advanced vehicle concepts, ULSAB-AVC	1999–2002	33 steel-producing companies in the world took part in the consortium Achieving CO_2 emissions at the level of 140 g/km Due to the vehicle's design, its weight can be reduced by 20–30%
Future steel vehicle, FSV	2008–2020	10 steel-producing companies in the world participated in the consortium It is planned to reduce the weight of the car by 35% for such vehicles as electric vehicles (EV), fuel cell electric vehicles (FCEV), and hybrid electric vehicles (HEV)

Table 2.2 Classification of automotive steels according to mechanical properties

Group 1. Steels with low tensile strength (less than 350 MPa)	Group 2. Steels with an average strength limit of break (350–500 MPa)	Group 3. Steels with a high tensile strength (500–800 MPa)	Group 4. Steels with ultra-high tensile strength (more than 800 MPa)
Mild 140YS/270TS BH 180YS/300TS BH 210YS/320TS BH 240YS/340TS	BH 260YS/370TS HSLA 280YS/350TS HSLA 350YS/450TS DP 300YS/500TS	DP 350YS/600TS TRIP 350YS/600TS DP 500YS/800TS TRIP 500YS/800TS CP 700YS/800TS	DP 700YS/1000TS MS 950YS/1200TS MS 1150YS/1400TS MS 1250YS/1520TS HS 950YS/1300TS

Note Steels of groups 3 and 4 belong to advanced high-strength steels (advanced high-strength steels, AHSS); TS is the ultimate tensile strength

One is a metallurgical designation providing some process information. Common designations include low-strength steels (interstitial-free and mild steels); conventional HSS, such as bake hardenable and high-strength, low-alloy steels (HSLA); and advanced high-strength steels, or AHSS (for example, dual phase and transformation-induced plasticity steels). Additional higher-strength steels for the automotive market include hot-formed, post-forming heat-treated steels designed for unique applications with improved edge stretch and stretch bending characteristics.

A second classification method important to part designers is the strength of the steel. We will use the general terms HSLA and AHSS to designate all higher-strength steels. This classification system has a problem with the ongoing development of the many new grades for each type of steel. Therefore, a DP or TRIP steel can have strength grades encompassing two or more strength ranges.

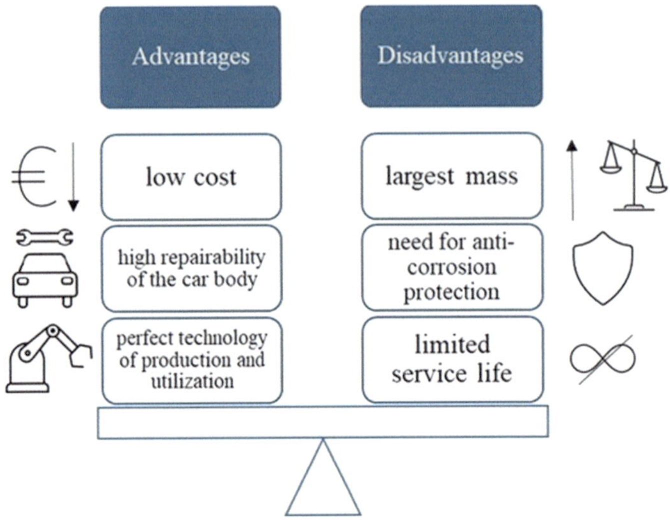

Fig. 2.2 Advantages and disadvantages of using steel in cars

A third classification method presents different steels' mechanical properties or forming parameters, such as total elongation, work hardening exponent (n-value), or hole expansion ratio. Figure 2.3 compares total elongations (a steel property related to formability) to the tensile strength of the current types of steel. These properties are important for press shop operations and virtual forming analyses.

The principal difference between conventional HSLA steels and AHSS is their microstructure [8]. Conventional HSLA steels are single-phase ferritic steels with a

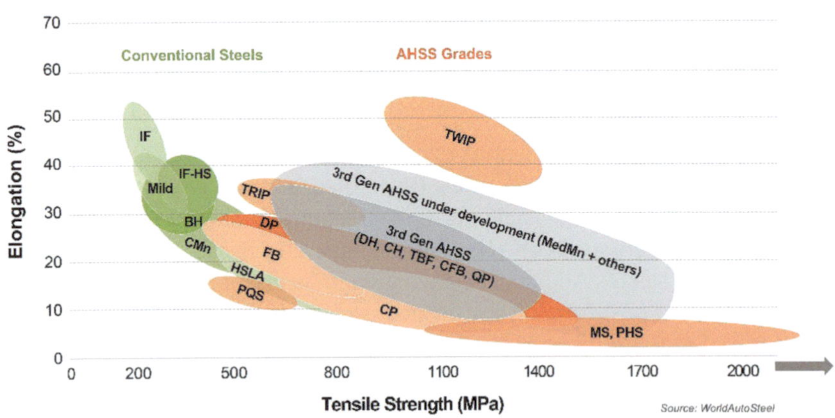

Fig. 2.3 The 2021 global formability diagram comparing strength and elongation of current and emerging steel grades [7] [Courtesy of WorldAutoSteel]

potential for some pearlite in C-Mn steels. AHSS are primarily steels with a multiphase microstructure containing one or more phases other than ferrite, pearlite, or cementite—for example, martensite, bainite, austenite, and/or retained austenite in quantities sufficient to produce unique mechanical properties. Some types of AHSS have a higher strain hardening capacity, resulting in a strength-ductility balance superior to that of conventional steels. Other types have ultra-high yield and tensile strengths and show a bake-hardening behavior.

2.2.1 *Typical Steels Used in the Automotive Industry*

Stainless steel. A steel raw material or pre-fabricated product can only have the «stainless» designation if it contains at least 10.5% chromium and 1% carbon alongside other alloying compounds. One striking feature of this steel type is its ability to resist corrosive materials. Its lustrous and light-bodied design has been synonymous with durability and functionality in many industries, not just vehicle manufacturing.

Advanced high-strength steel. Modern steel innovations have led to a new grade of advanced-high-strength steel or AHSS. Unlike the typical low-carbon and mild steel, AHSSs are sophisticated and complex. It's made out of carefully selected materials with specific chemical compositions. The result is a precisely heated steel product that is tough, ductile, and has an exceptional fatigue rating.

Steels with high-carbon content, known as high-carbon steel, are alloys with a heat treatment method known as quenching. In traditional steel heat treatment, wherein the carbon content is dissolved into the heated iron, rapid cooling is applied to quenched steel, which "traps" the carbon and transforms the steel's structuring.

Low-carbon steel. Low-carbon steel has a much lower tensile strength and poorer yield-point runout than its high-carbon counterpart. Also referred to as mild steel, low-carbon steel has continued to find its way into many vehicle parts because it is cheap to produce and useful for producing smaller components.

Galvanized steel. Cheaper than stainless steel but more effective than aluminum, galvanized steel is another popular alloyed metal used in vehicle-making. Today, galvanized steel is even considered a standard in any vehicle. Almost 80% of cars, trucks, and motorcycles are galvanized steel.

2.2.2 *Types of Progressive Steels*

Bake hardenable steels (BH) chemically containing max 0.06–0.11% C, max 0.50% Si, max 0.7–1.0% Mn, max 0.060–0.120% P, max 0.030% S, max 0.015% Al, max 0.12% Ti, max 0.09% Nb. BH-steel flat products are used to manufacture car bodies, which require a combination of easy formability and high strength. A special feature of BH steel is that heat treatment is carried out when applying paint/galvanization.

High-strength low-alloy steels (HSLA) chemically contain from 0.10 to 0.26% C, from 0.5 to 1.65% Mn, and Nb, Zr, Mo, V, and Ti. According to ASTM classification, these are steels A 572, A 588, A 606, A 607, etc. The properties of these steels are determined by using different microalloying options and controlled rolling and accelerated cooling. It should be noted that the price for rolling HSLA steel is based on carbon steel, not alloy steel. Rolled HSLA steels have many applications, from pipelines to construction and agricultural machinery, components to passenger cars, and bridge structures.

Dual-phase steel (DP) chemically contains 0.10–0.17% C, 0.8–1.5% Mn, 0.2–0.5% Si, 0–0.015% Nb, 0– 0.002% B. DP steels have a microstructure consisting of hard martensite particles in a soft and ductile ferrite matrix. By thickness, DP-steel rolls are divided into three groups: less than 1.5 mm, 2–2.5 mm, and more than 2.5 mm. The last assortment in the hot-rolled version is used for the car wheel disc. The hot-rolled medium group is used to produce truck bumpers and chassis.

Transformation-induced plasticity steel (TRIP) chemically contains 0.14–0.19% C, 1.57–1.58% Mn, 02–1.6% Si, 0.013–0.018% P, 0.025–0.028% S, 0.04–0.08% Cr, 0.02% Ni, 0.002% Cu, 0.036–0.043% Al, 0.0–0.027% Ti, 0.0–0.042% Nb. The microstructure of TRIP steels consists of austenite embedded in the primary matrix of ferrite, as well as martensite and bainite. Typical flat rolled thicknesses of TRIP steels are 2–5 mm for hot-rolled grades and 0.6–1.5 mm for cold-rolled grades. These steels are used for complex automotive parts (car body panels, beams, etc.) where the best formability is required, and the part cannot be made from DP or CP steels.

Martensitic steel (MS) chemically contains 0.12–0.26% C, max 1.4% Si, 0.5–2.1% Mn, Ti, Nb, V, Cr, Mo, B, Cu, Ni. MS-steels are characterized by a martensitic matrix containing a small amount of ferrite and/or bainite. In the group of multiphase steels, MS steels have the highest level of tensile strength. Transverse beams, as well as beams and bumper fittings, are made from this steel.

Complex phase steel (CP) has the following chemical composition: max 0.18–0.23% C, max 0.80–1.00% Si, max 2.20–2.70% Mn, max 0.080% P, max 0.015% S, 0.015–2.0% Al, max 1.00% Cr + Mo, max 0.15% Ti + Nb, max 0.20% V, max 0.005% B (EN 10,346 and/or EN 10,338). High-strength CP steel is characterized by a multiphase microstructure consisting of a thin ferritic matrix and solid inclusions of bainite, martensite, and a small amount of austenite. These steels combine high strength and plasticity with high energy absorption, which is why they are used to manufacture bumpers and other similar auto parts.

2.3 Advanced High-Strength Steels

All advanced high-strength steels (AHSS) are produced by controlling the chemistry and cooling rate from the austenite or austenite plus ferrite phase, either on the runout table of the hot mill (for hot-rolled products) or in the cooling section of the continuous annealing furnace (continuously annealed or hot-dip coated products).

Research has provided chemical and processing combinations that have created many additional grades and improved properties within each type of AHSS.

2.3.1 Dual-Phase Steel

Dual-phase (DP) steels consist of a ferritic matrix containing a hard martensitic second phase in the form of islands [9]. Figure 2.4 shows a microstructure of DP steel, which contains ferrite plus islands of martensite. The soft ferrite phase is generally continuous, giving these steels excellent ductility. When these steels deform, strain is concentrated in the lower-strength ferrite phase surrounding the islands of martensite, creating the unique high initial work-hardening rate (n-value) exhibited by these steels.

Increasing the volume fraction of hard secondary phases typically enhances strength. Dual-phase (DP) steels are created through controlled cooling from the austenite phase in hot-rolled products or from the two-phase ferrite plus austenite phase in continuously annealed cold-rolled and hot-dip coated products. This process transforms some austenite into ferrite, followed by rapid cooling that converts the remaining austenite into martensite. Due to the production method, various other phases, such as bainite and retained austenite, may also be present. Depending on the composition and processing route, steels that require improved resistance to cracking at stretched edges—often assessed by hole expansion capacity—can exhibit a microstructure with significant amounts of bainite.

The work hardening rate and outstanding elongation of DP steels result in ultimate tensile strengths that are significantly greater than those of conventional steels with comparable yield strength [11]. Figure 2.5 compares the engineering stress–strain curves for HSLA and DP steel with similar yield strength.

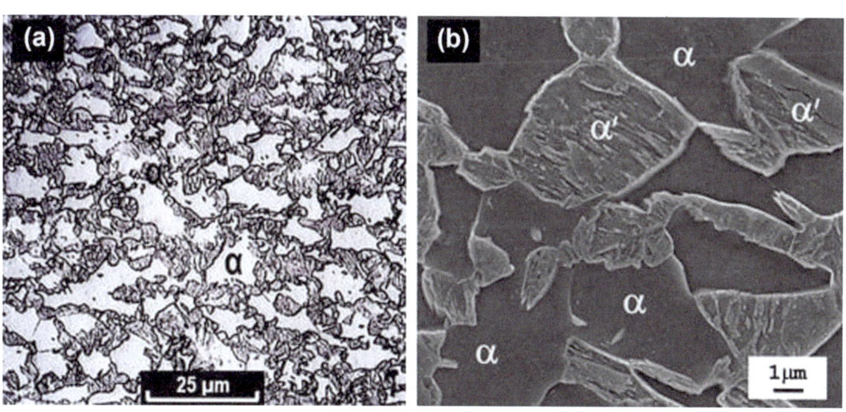

Fig. 2.4 **a** Base metal microstructure DP980, **b** phase detail (α = ferrite, α' = martensite) [10]

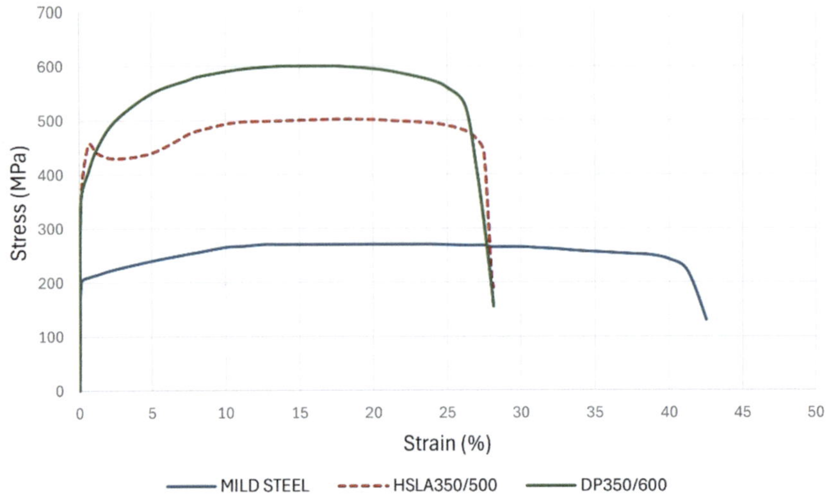

Fig. 2.5 Stress–strain curves for DP350/600

The DP steel demonstrates a higher initial work hardening rate, greater ultimate tensile strength, and lower yield strength to ultimate tensile strength ratio compared to the HSLA steel with equivalent yield strength. Engineering and true stress–strain curves for different grades of DP steels are presented in Fig. 2.6.

DP and other AHSS steels exhibit a bake-hardening effect that significantly enhances the performance of conventional higher-strength steels. This bake-hardening effect refers to the increase in yield strength that occurs due to elevated temperature aging, induced by the curing temperature of paint bake ovens, following pre-straining, which results from work hardening during stamping or other manufacturing processes. The degree of the bake-hardening effect in AHSS is influenced by the appropriate level of forming strain tailored to the specific chemistry and thermal history of the steel.

Examples of current production grades of DP steels and typical automotive applications are shown in Table 2.3.

Each steel grade is identified by metallurgical type, minimum yield strength (in MPa), and minimum tensile strength (in MPa). For example, DP 500/800 means dual-phase steel with 500 MPa minimum yield strength and 800 MPa minimum ultimate tensile strength.

2.3.2 *Transformation-Induced Plasticity Steel*

The microstructure of transformation-induced plasticity (TRIP) steels is retained austenite embedded in a primary ferrite matrix. In addition to a minimum of 5 vol.% of retained austenite, hard phases such as martensite and bainite are present in varying

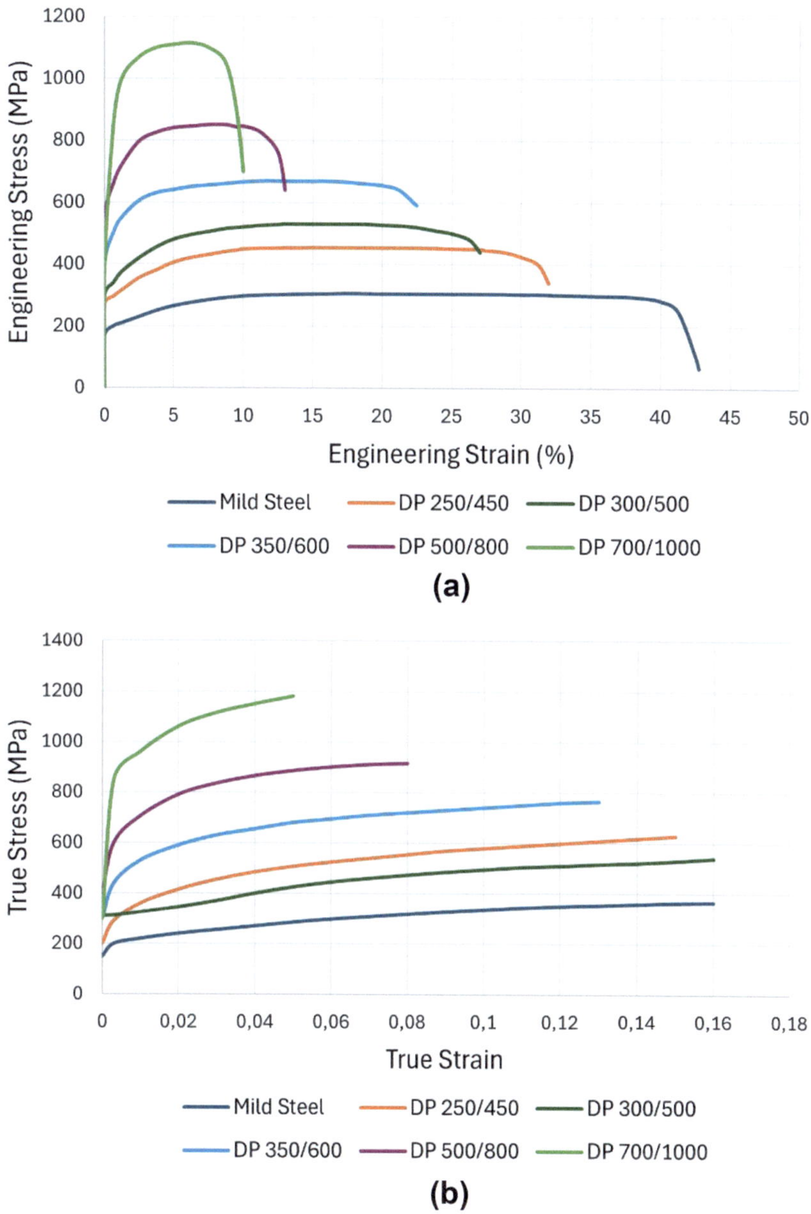

Fig. 2.6 **a** Engineering stress–strain and **b** true stress–strain curves of DP steel grades

Table 2.3 DP steels and typical automotive applications

Grade	Application
DP 300/500	Roof outer, door outer, body side outer, package tray, floor panel
DP 350/600	Floor panel, hood outer, body side outer, cowl, fender, floor reinforcements
DP 500/800	Body side inner, quarter panel inner, rear rails, rear shock reinforcements
DP 600/980	Safety cage components (B-pillar, floor panel tunnel, engine cradle, front sub-frame package tray, shotgun, seat)
DP 700/1000	Roof rails
DP 800/1180	B-pillar upper

amounts [12]. TRIP steels typically require an isothermal hold at an intermediate temperature, which produces some bainite. The higher silicon and carbon content of TRIP steels also result in significant volume fractions of retained austenite in the final microstructure [13]. The microstructures of the cold rolled, and three TRIP-aided steels are shown in Fig. 2.7.

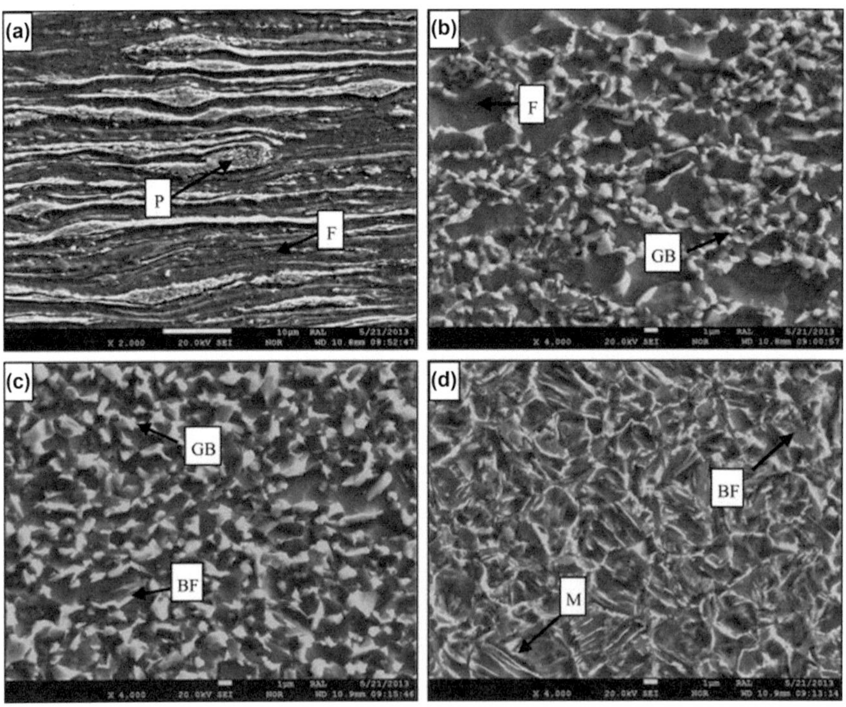

Fig. 2.7 Microstructural evolution using SEM of different types of TRIP-steel: **a** cold-rolled, **b** ferritic matrix (F-TRIP), **c** bainitic matrix (B-TRIP) and **d** martensitic matrix (M-TRIP) (F—ferrite, P—pearlite, GB—granular bainite, BF—bainitic ferrite, M—martensite) [14]

The results of the tensile tests for every microstructure are presented s engineering stress-engineering strain curves in Fig. 2.8. Significant differences were noted in the work-hardening behaviors, which can be directly attributed to the stability of retained austenite in the various microstructures. Uniaxial tensile tests showed that the M-TRIP sample demonstrates greater strength and a higher work hardening rate at low strains, followed by a rapid decline at high strains [14].

Figure 2.9 compares the engineering stress–strain curve for HSLA steel to a DP steel curve of similar yield strength. The yield point elongation in TRIP steel is related to a low density of dislocations in ferrite grains [15].

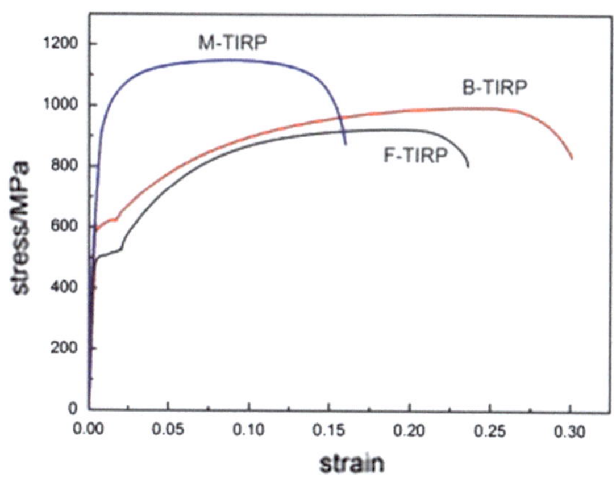

Fig. 2.8 Tensile stress–strain curves of different types of TRIP-steel: ferritic matrix (F-TRIP), bainitic matrix (B-TRIP) and martensitic matrix (M-TRIP) [14]

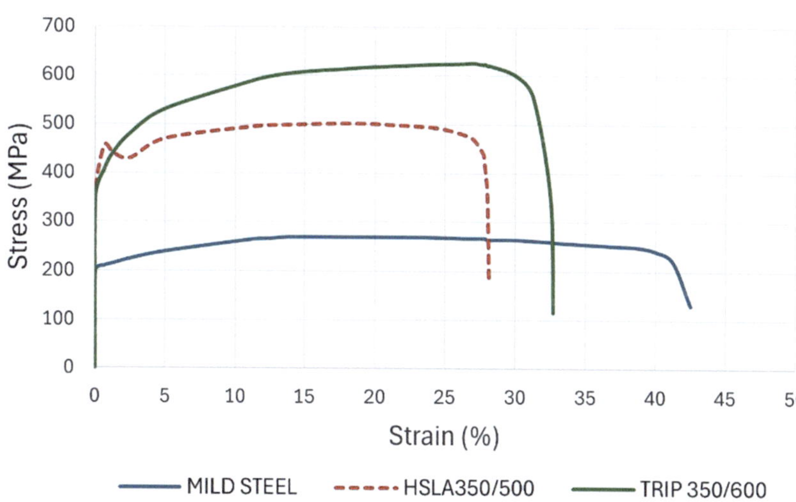

Fig. 2.9 Stress–strain curves for TRIP350/600

During deformation, the distribution of hard second phases within soft ferrite results in a high work hardening rate, as seen in DP steels. In contrast, TRIP steels experience a gradual transformation of retained austenite into martensite as strain increases, which enhances the work hardening rate at elevated strain levels [11]. This phenomenon is showed in Fig. 2.10, which compares the engineering stress–strain behavior of HSLA, DP, and TRIP steels with approximately similar yield strengths. While the TRIP steel exhibits a lower initial work hardening rate compared to the DP steel, its hardening rate continues to be significant at higher strains, whereas the work hardening of the DP steel begins to decline. Engineering and true stress–strain curves for different grades of TRIP steels are presented in Fig. 2.11.

The work-hardening rates of TRIP steels are significantly higher than those of conventional high-strength steels (HSS), enabling substantial stretch-forming capabilities. This characteristic is particularly beneficial for designers who leverage the elevated work hardening rate and enhanced bake hardening effect to create components that utilize the mechanical properties achieved during forming. TRIP steels maintain a high work-hardening rate even at greater strains, offering a slight advantage over DP steels in the most demanding stretch-forming scenarios. To stabilize the retained austenite phase below ambient temperature, TRIP steels incorporate higher levels of carbon compared to DP steels. Additionally, increased levels of silicon and/or aluminum promote the formation of ferrite and bainite. These elements help preserve the necessary carbon content within the retained austenite. TRIP steels need to suppress carbide precipitation during the bainitic transformation, and silicon and aluminum are employed to prevent such precipitation in the bainite region.

Examples of current production grades of TRIP steels and typical automotive applications are shown in Table 2.4.

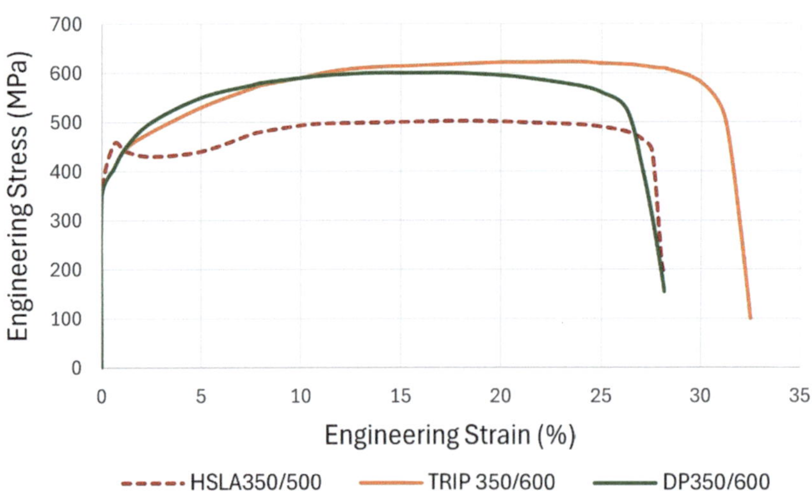

Fig. 2.10 Engineering stress–strain curves of HSLA, DP, and TRIP steels

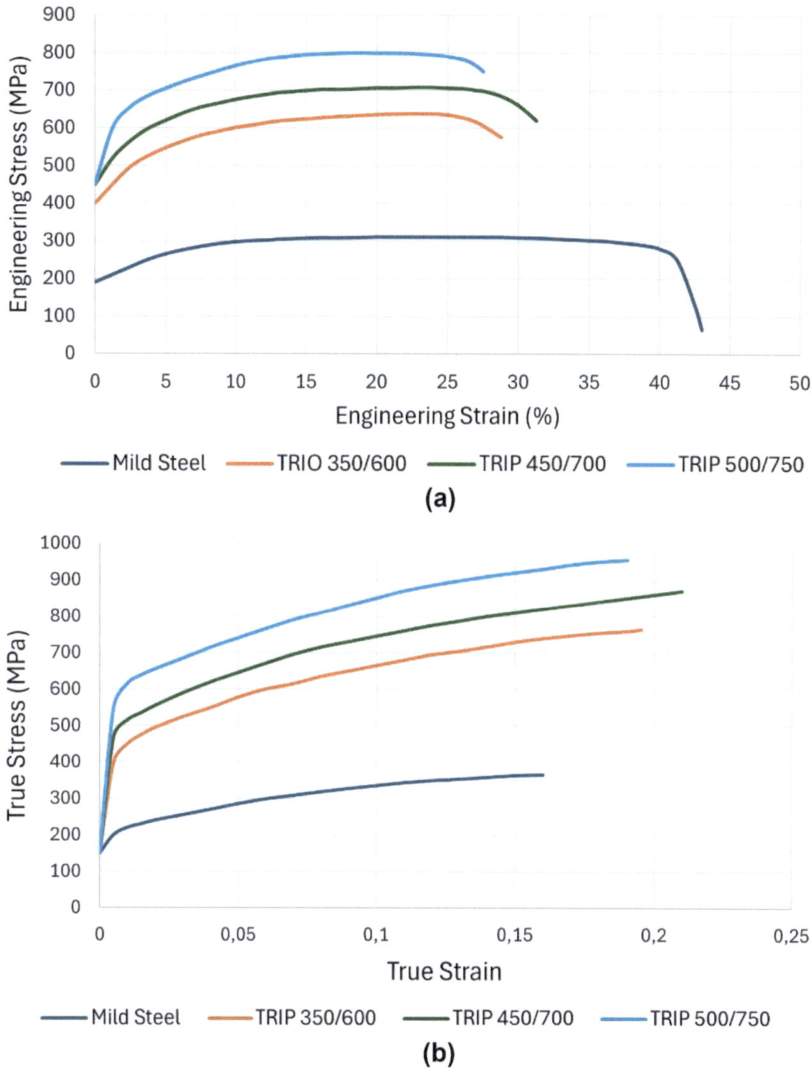

Fig. 2.11 **a** Engineering stress–strain and **b** true stress–strain curves of TRIP steel grades

Table 2.4 TRIP steels and typical automotive applications

Grade	Application
TRIP 600	Inner door panels, crash boxes, structural reinforcements
TRIP 700	B-pillars, rocker reinforcements, chassis components
TRIP 800	Side impact beams, A-pillars, underbody reinforcements
TRIP 900	Roof rails, frame reinforcements, intrusion beams
TRIP 980	Front and rear rail reinforcements, bumper beams

TRIP steels are marked based on their minimum tensile strength. The naming convention typically follows the format TRIP XXX, where XXX represents the tensile strength in MPa.

2.3.3 Complex Phase Steel

Complex phase (CP) steels typify the transition to steel with very high ultimate tensile strengths. The microstructure of CP steel contains small amounts of martensite, retained austenite, and pearlite within the ferrite/bainite matrix. Extreme grain refinement is created by retarded recrystallization or precipitation of microalloying elements like Ti or Nb. The microstructure of complex-phase steel is shown in Fig. 2.12. SEM images showed the ferritic matrix with significant numbers of martensitic/austenitic islands and bainitic ferrite regions.

Particularly, bainite could smoothen the property gradient between soft ferrite and hard martensite phases. Thus, CP steel shows superior formability, hole expansion tolerance, and higher yield strength than other AHSS steels with similar ultimate tensile strength [17].

Figure 2.13 compares the engineering stress–strain curve for HSLA steel to a CP 1000/1200 steel curve. Compared to DP steels, CP steels show significantly higher yield strengths at equal tensile strengths of 800 MPa and greater [11]. CP steels have high energy absorption, high residual deformation capacity, and good hole expansion. Engineering and true stress–strain curves for CP steel grades are presented in Fig. 2.14.

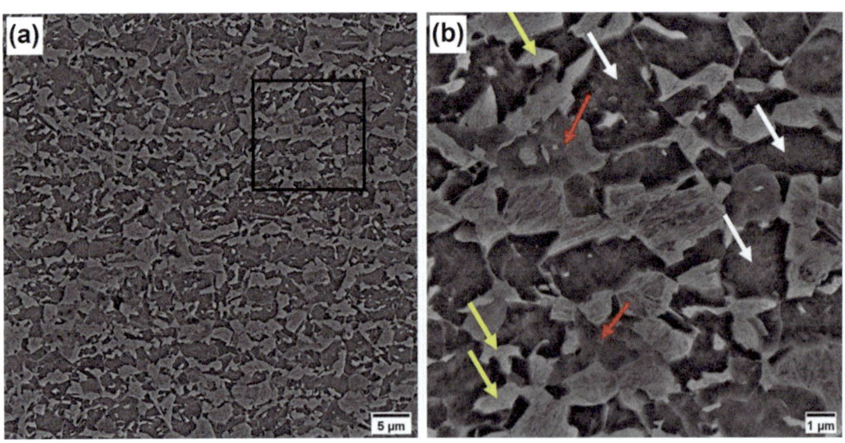

Fig. 2.12 SEM images of the CP1100 steel: **a**—deeply etched ferritic matrix (dark) with martensite/retained austenite islands, **b**—higher magnification image yellow arrows show the M/A islands, white arrows the ferrite and red arrows some bainitic ferrite [16]

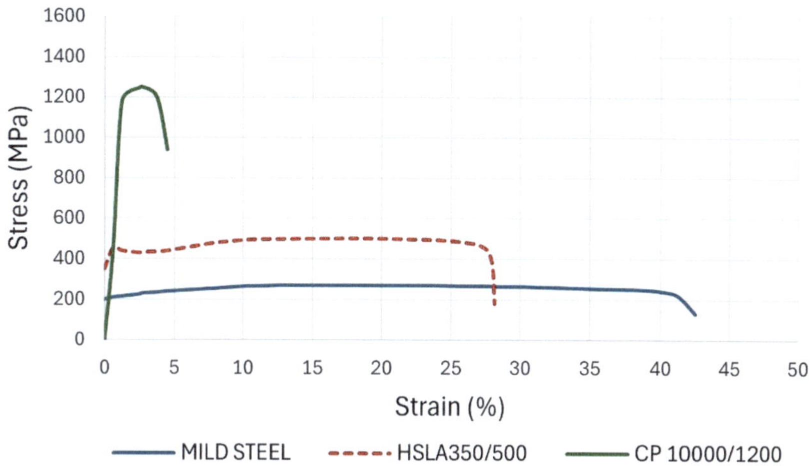

Fig. 2.13 Stress–strain curves for CP 1000/1200

At present, CP steel is widely used to manufacture structural parts of automobiles, such as torsion beams, control arms, and slide rails. CP steel could also be used to manufacture automotive panels after proper bake hardening [17]. Current production grades of CP steels and example automotive applications are shown in Table 2.5.

2.3.4 *Martensitic Steel*

To create martensitic steels (MS), the austenite that exists during hot-rolling or annealing is transformed almost entirely to martensite during quenching on the run-out table or in the cooling section of the continuous annealing line. The MS steels are characterized by a martensitic matrix containing small amounts of ferrite and/or bainite (Fig. 2.15).

In multiphase steels, MS steels show the highest tensile strength level. This structure also can be developed with post-forming heat treatment. MS steels provide the highest strength, up to 1700 MPa ultimate tensile strength [11]. MS steels are often subjected to post-quench tempering to improve ductility and can provide adequate formability even at extremely high strengths. Figure 2.16 shows MS950/1200 compared to HSLA. Engineering and true stress–strain curves for MS steel grades are presented in Fig. 2.17.

Current production grades of MS steels and example automotive applications are shown in Table 2.6.

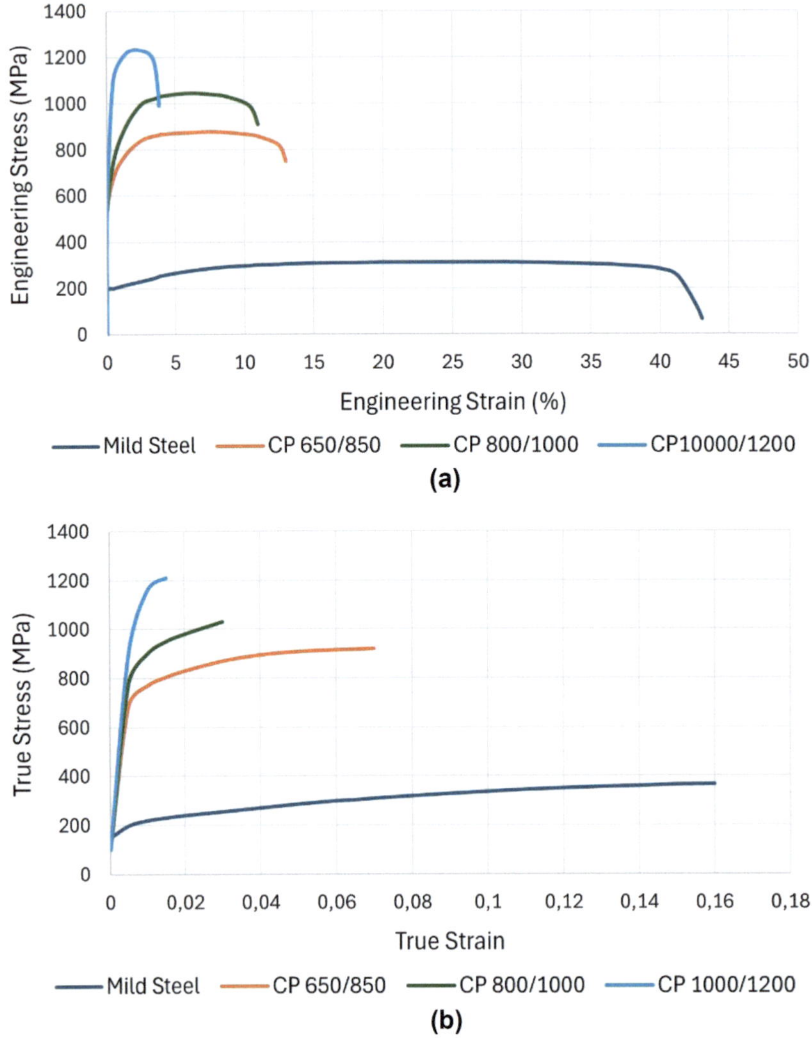

Fig. 2.14 **a** Engineering stress–strain and **b** true stress–strain curves of CP steel grades

Table 2.5 CP steels and typical automotive applications

Grade	Application
CP 600/900	Frame rails, B-pillar reinforcements
CP 680/780	Frame rails, chassis components, transverse beams
DP 750/900	B-pillar reinforcements, tunnel stiffener
DP 800/1000	Rear suspension brackets, fender beam
DP 1000/1200	Rear frame rail reinforcements, rocker outer
DP 1050/1470	Rocker panels, bumper beams

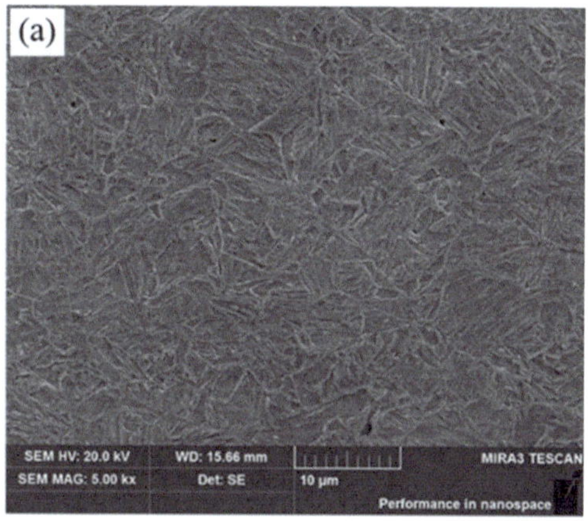

Fig. 2.15 SEM micrographs of MS1400 [18]

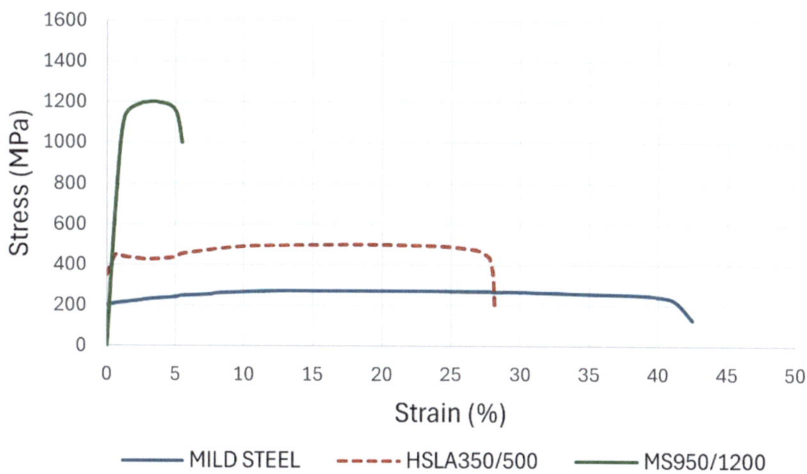

Fig. 2.16 Stress–strain curves for MS950/1200

2.3.5 *Ferritic-Bainitic Steel*

Ferritic-bainitic (FB) steels are sometimes utilized to meet specific customer application requirements that require Stretch Flangeable (SF) or High Hole Expansion (HHE) capabilities for improved edge stretch capability [11]. FB steels are available as hot-rolled products.

FB steels have a microstructure of fine ferrite and bainite (Fig. 2.18). Strengthening is obtained by both grain refinement and second-phase hardening with bainite.

The primary advantage of FB steels over HSLA and DP steels is the improved stretchability of sheared edges as measured by the hole expansion test (λ). Compared

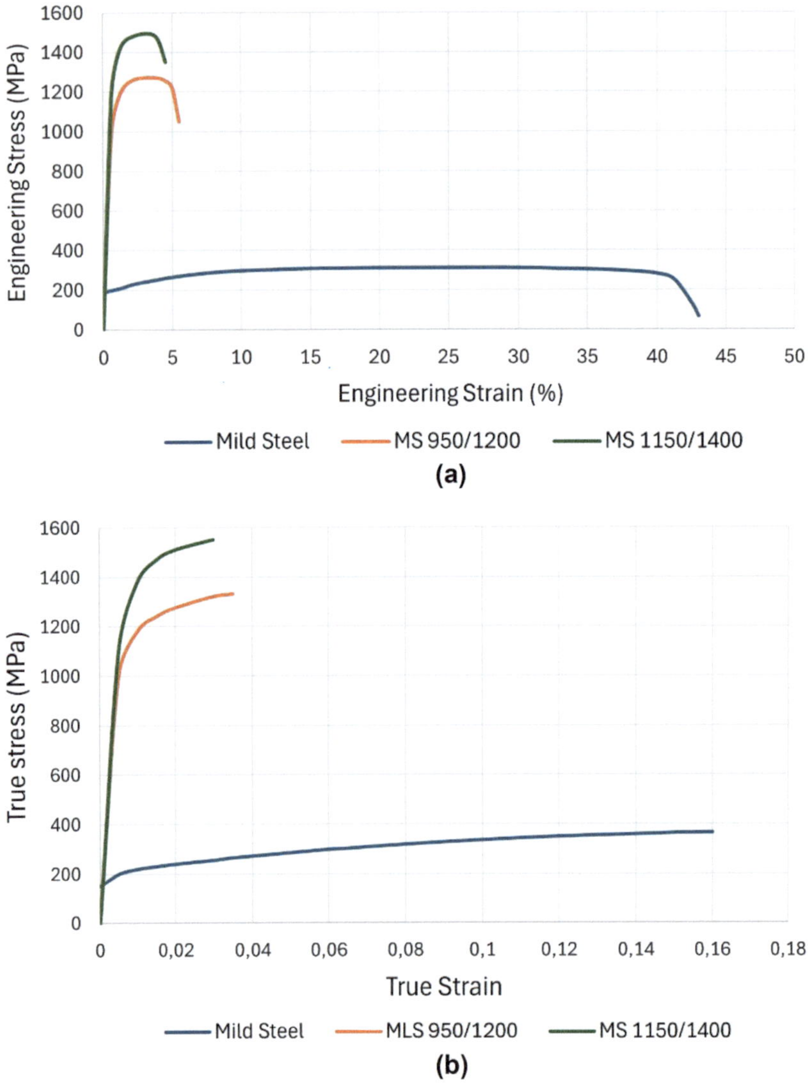

Fig. 2.17 **a** Engineering stress–strain and **b** true stress–strain curves of MS steel grades

Table 2.6 MS steels and typical automotive applications

Grade	Application
MS 950/1200	Cross-members, side intrusion beams, bumper beams, bumper reinforcements
MS 1150/1400	Rocker outer, side intrusion beams, bumper beams, bumper reinforcements
MS 1250/1500	Side intrusion beams, bumper beams, bumper reinforcements

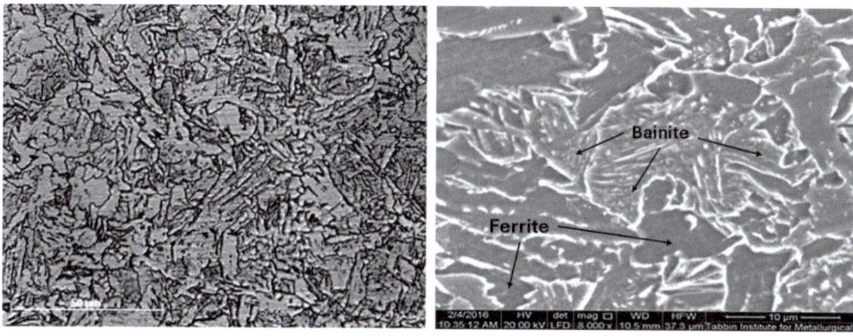

Fig. 2.18 SEM micrograph of the experimental ferrite bainite steel T-A (content C 0.168%, Mn 0.83%, V 0.015%) [19]

to HSLA steels with the same level of strength, FB steels also have a higher strain hardening exponent (n-value) and increased total elongation. Figure 2.19 compares FB 450/600 with HSLA 350/500 steel. Engineering and true stress–strain curves for FB steel are shown in Fig. 2.20.

Because of their good weldability, FB steels are considered for tailored blank applications. Both good crash performances and good fatigue properties also characterize these steels.

Current production grades of FB steels and example automotive applications are shown in Table 2.7.

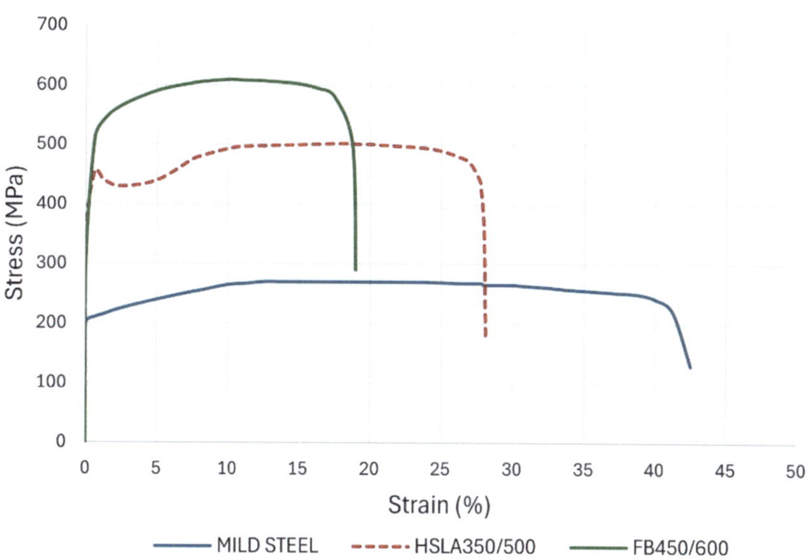

Fig. 2.19 Stress–strain curves for FB450/600

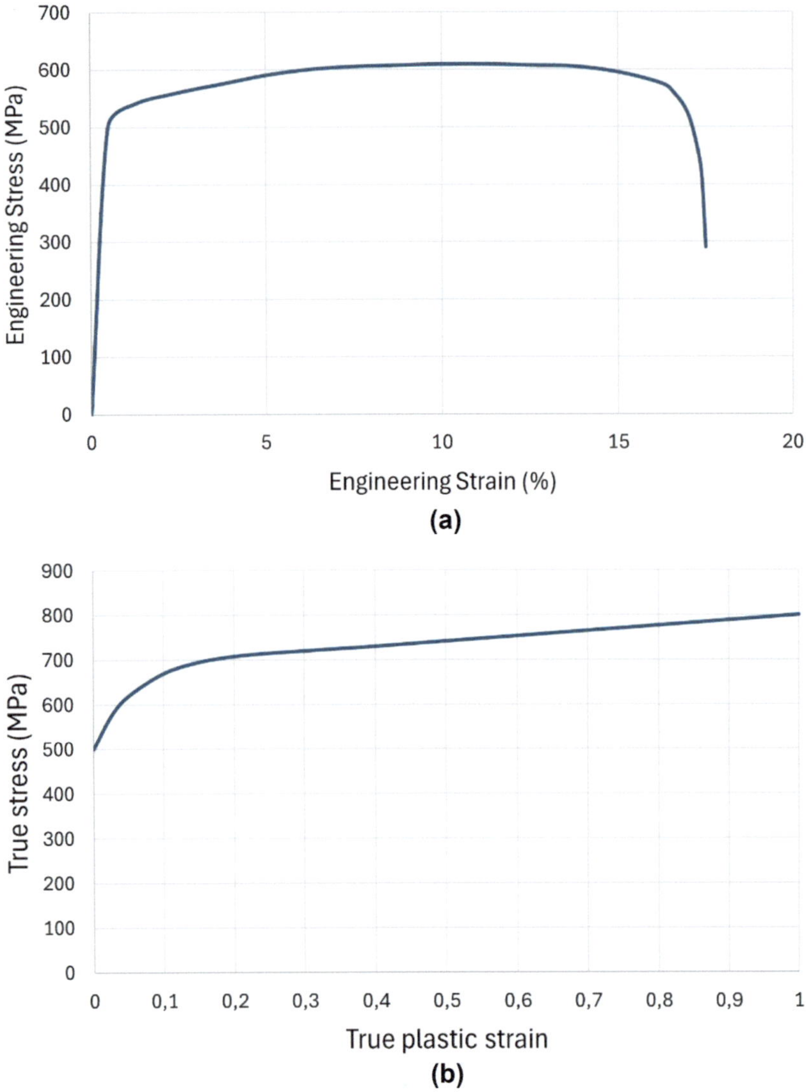

Fig. 2.20 Engineering stress–strain (**a**) and true stress–strain (**b**) curve for FB 450/600

Table 2.7 FB steels and typical automotive applications

Grade	Application
FB 330/450	Rim, brake pedal arm, seat cross member, suspension arm
FB 450/600	Lower control arm, rim, bumper beam, chassis parts, rear twist beam

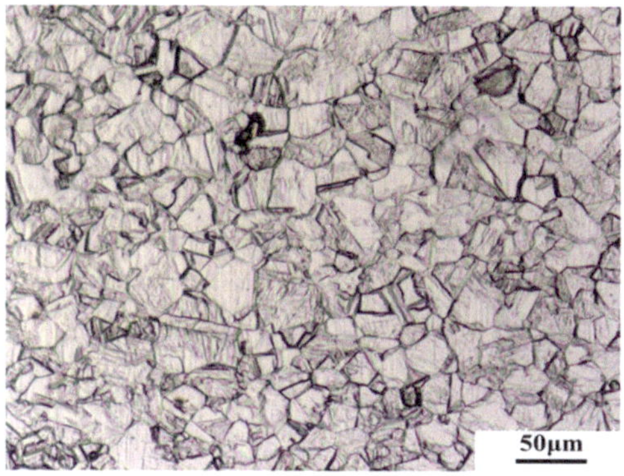

Fig. 2.21 Microstructure of the Fe-Mn-Al-C TWIP steel [21]

2.3.6 Twinning-Induced Plasticity Steel

Twinning-induced plasticity (TWIP) steels have a high manganese content (17–24%) that causes the steel to be fully austenitic at room temperatures. The formation of deformation twins drives a large amount of deformation. This deformation mode leads to the naming of this steel class. The twinning causes a high instantaneous hardening rate (n-value) value as the microstructure becomes finer and finer. The resultant twin boundaries act like grain boundaries and strengthen the steel [20].

Figure 2.21 shows the annealed microstructure for TWIP steel.

TWIP steels combine extremely high strength with extremely high stretchability. The n-value increases to a value of 0.4 at an approximate engineering strain of 30% and then remains constant until both uniform and total elongation reaches 50%. The tensile strength is higher than 1000 MPa [11].

Figure 2.22 compares TWIP 750/1000 to HSLA steel.

Engineering and true stress–strain curves for TWIP steel grades are shown in Fig. 2.23.

TWIP steels have high strength and elongation percentages but are much more expensive than others (Table 2.8).

Current production grades of TWIP steels and example automotive applications are shown in Table 2.9.

2.3.7 Hot-Formed Steel

Implementing press-hardening applications and using hardenable steels are promising alternatives for optimized part geometries with complex shapes and no spring back issues. Boron-based hot-forming steels (between 0.001 and 0.005% B) have been used since the 1990s for body-in-white construction [23]. A typical

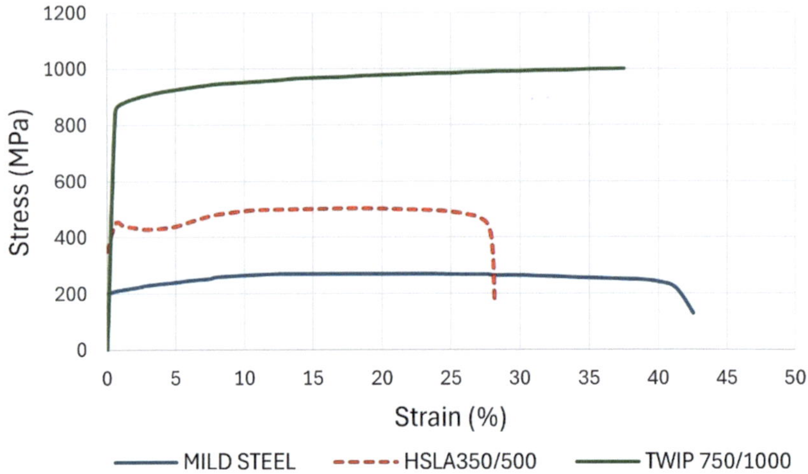

Fig. 2.22 Stress–strain curves for TWIP750/1000

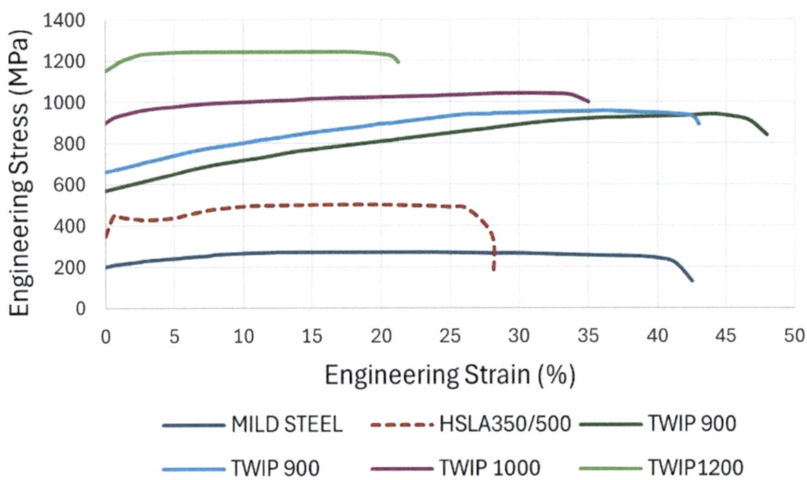

Fig. 2.23 Engineering stress–strain curve for TWIP steel

minimum temperature of 850 °C must be maintained during the forming process (austenitization), followed by a cooling rate greater than 50 °C/s to achieve the desired mechanical properties [11].

Two types of press-hardening or hot-forming applications are currently available [24]:

- Direct hot-forming,
- Indirect hot-forming.

Table 2.8 Tensile properties of TWIP steel compared with HSLA, DP, and TRIP steels (in the transverse direction) [22]

Steel	Yield strength (MPa)	Tensile strength (MPa)	Uniform elongation (%)	Total elongation (%)	Strain hardening exponent (n)
HSLA 320	342	428	16	24	0.169
DP 590	342	629	16	26	0.152
DP 980	644	1009	10	15	0.112
TRIP 800	478	825	21	27	0.216
TWIP 1000	496	1102	52	52	0.415

Table 2.9 TWIP steels and typical automotive applications [21]

Grade of TWIP steel	Automotive application
TWIP 500/900	A-pillar, wheelhouse, front side member
TWIP 500/980	Wheel, lower control arm, front and rear bumper beams, B-pillar, wheel rim
TWIP 600/900	Floor cross-member, wheelhouse
TWIP 750/1000	Door impact beam
TWIP 950/1200	Door impact beam

All blank deformation is done in the high-temperature austenitic range during *direct hot-forming*, followed by quenching. *Indirect hot-forming* performs the blank at room temperature to a high percentage of the final part shape, followed by additional high-temperature forming and quenching.

The final microstructure of hot-formed (HF) steel is similar to martensite. Stress–strain curves after quenching are similar to martensitic steels shown in Fig. 2.15. Current production grades of HF steels and automotive applications are shown in Table 2.10.

Table 2.10 HF steels and typical automotive applications

Grade	Automotive application
HF 340/480	As-received room temperature
HF 1050/1500	Heat treated after forming A-pillar, B-pillar, cross beam
HF 1200/1900	Heat treated after forming

Table 2.11 PFHT steels and typical automotive applications

Grade	Automotive application
PFHT 340/480	As-received room temperature
PFHT 1050/1500	Heat treated after forming
PFHT 1200/1900	Heat treated after forming

2.3.8 *Post-forming Heat-Treatable Steel*

Post-forming heat treatment is a general method to develop an alternative to higher-strength steel. The major issue holding back the widespread implementation of HSS typically has been maintaining part geometry during and after the heat treatment [11]. To address distortion problems, the part is fixtured and heated (via furnace or induction), followed by immediate quenching. This has been a more effective solution for production applications because the stamping is formed at a lower level and then raised to a much higher strength by heat treatment.

Another process is the air-hardening of alloyed tempering steels that feature very good forming properties in the soft state (deep-drawing properties) and high strength after heat treatment (air-hardening) [25]. Apart from direct application as sheet material, air-hardening steels are suitable for tube welding. These tubes are excellent for hydroforming applications.

The third option is in-die quenching [26]. A version of Indirect Hot-Forming completes all forming of the part at room temperature, heats the part to about 850–900 °C, and then uses a water-cooled die to quench the part to martensite. This process is called form hardening.

Current production grades of PFHT steel and example automotive applications are shown in Table 2.11.

2.3.9 *Specially Processed Steels*

DP, TRIP, and other AHSS microstructures have islands or particles in different phases. One example is the martensite islands. When forming steels, these particles cause strain localization during edge stretching, stretch-bending over tight radii, or other deformation modes whereby strain can localize to generate high peak strains. These steels are then subject to failure at deformations that are less than predicted by the properties. Steel companies have been conducting extensive research to develop specialized steels for applications that minimize the negative effects of these phases or particles.

Table 2.12 The nanoprecipitation grades and typical automotive applications

Grade	Application
TPN 680/780: HR660Y760T-TP	B-pillar, crossmember, longitudinal members, side panel reinforcement
TPN 750/900: HR730Y880T-TP	B-pillar, crossmember, longitudinal members, bumper, side panel reinforcement

2.3.10 Evolving Advanced High-Strength Steels Types

In response to automotive demands for additional AHSS capabilities, research laboratories in the steel industry and academic institutions continue to search for new types of steel or modifications of existing types. The primary research goal (enhanced multiphase products) is improving formability for a given strength range while reducing the cost and welding problems associated with steels having a high percentage of austenite. Other developing steels are ultrafine (nano) grain, low density, and high Young's modulus steels.

2.3.10.1 Three-Phase Steel with Nanoprecipitation

Three-phase steel with nanoprecipitation (TPN) is a new family of high-performance cold-forming steels [11]. It consists of a ferritic matrix containing bainite, residual austenite, and nanoprecipitations. A high percentage of retained austenite is converted to martensite during the forming. This work-induced transformation positively affects the hardening behavior of the material to create a good combination of high formability and strength. Due to their high resistance to local thinning, these grades allow component geometries that are very difficult to achieve with other high-strength steels of the same strength class [27].

The nanoprecipitation grades produced for automotive applications are shown in Table 2.12.

2.3.10.2 Quenching and Partitioning

Quenching and partitioning (Q&P) steels are a series of C-Si-Mn, C-Si-Mn-Al, or other likely compositions subjected to the quenching and partitioning heat treatment. With a final microstructure of ferrite (in the case of partial austenitization), martensite, and retained austenite (Fig. 2.24), Q&P steels exhibit an excellent combination of strength and ductility, which permits their use in a new generation of AHSS for automobiles.

Q&P steels are suitable for cold stamping for those structure and safety parts of automobiles with relatively complicated shapes to improve fuel economy while promoting passenger safety.

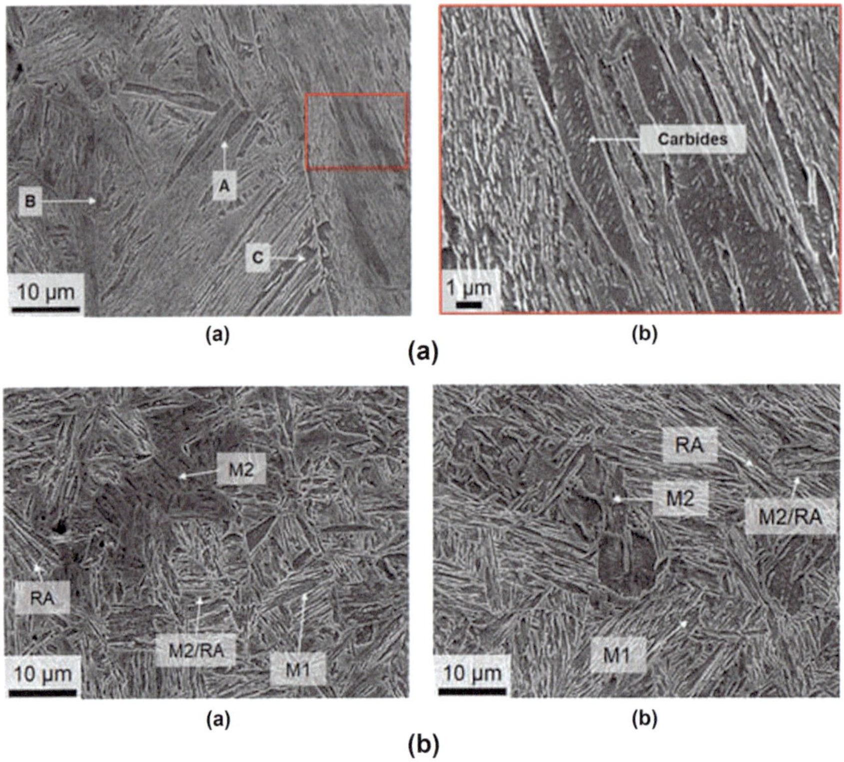

Fig. 2.24 SEM micrograph of the **a** as-received microstructure of the Q&P steel (0.2C-3.5Mn-1.5Si (wt%)), **b** after the QP process: M1 stands for tempered martensite, M2 for fresh martensite, and RA for retained austenite [28]

2.4 Conventional Low- and High-Strength Automotive Sheet Steels

Mild steels. Mild steels have an essentially ferritic microstructure. Drawing quality (DQ) and aluminum killed (AKDQ) steels are mild and often serve as a reference base because of their widespread application and large production volume over the past decades.

Interstitial-free (IF) steels (low strength and high strength). IF steels have ultra-low carbon levels designed for lower yield strengths and higher work hardening exponents (n-values). These steels have more stretchability than mild steels. The IF-HS grades utilize elements for solid solution strengthening, precipitation of carbides and/or nitrides, and grain refinement. Another common element added to increase strength is phosphorous (another solid solution strengthener). The higher strength grades of IF steel are widely used for structural and closure applications.

Bake hardenable (BH) steels. BH steels have a basic ferritic microstructure and solid solution strengthening. A unique feature of these steels is the chemistry and processing designed to keep carbon in solution during steelmaking and then allow this carbon to come out of solution during paint baking or several weeks at room temperature. This increases the yield strength of the formed part for increased dent resistance without a reduction in formability. Common applications are automotive outer body panels where increased dent resistance is required.

Carbon-manganese (CM) steels. CM steels utilize solid solution strengthening for higher strength.

High-strength low-alloy (HSLA) steels. HSLA steels increase strength primarily by micro-alloying elements contributing to fine carbide precipitation, substitutional and interstitial strengthening, and grain-size refinement. HSLA steels are found in many body-in-white and underbody structural applications where strength is needed for increased in-service loads.

2.5 Stainless Steels

The density and modulus of stainless steel are very close to the density and modulus of steel. Therefore, directly substituting steel with stainless steel does not reduce weight in stiffness-critical applications. In strength-critical applications, stainless steel can provide weight reduction over steel for the following reasons [29]:

- The yield strength-to-density ratio of several stainless steels is higher than that of high-strength steels.
- Stainless steel has a higher work hardening coefficient and formability than steel, which means it can tolerate higher uniform plastic deformation and thickness reduction during forming.
- Stainless steel has a higher strain rate sensitivity than steel, which can absorb higher crash energy than steel. Additionally, it has the capability of collapsing progressively in a controlled manner.

Another great advantage of stainless steel is its corrosion resistance. Anticorrosion coatings may not be needed if stainless steel is used instead of steel. Despite the above advantages, stainless steel has found very little application in automotive structures because of its high cost. A few structural applications where stainless steel has been tried are fuel tanks, knuckle arms, and wheels [30].

Stainless steel is available in a variety of grades. Still, the two grades used for automotive applications are the austenitic grade (300 series alloys, containing Cr and Ni as the principal alloying elements) and the ferritic grade (400 series alloys, containing Cr as the principal alloying element). The austenitic grade is non-magnetic and has higher yield strength, ductility, and corrosion resistance than the ferritic grade. Heat treatment can strengthen neither grade, but cold work can strengthen both grades. The austenitic grade has a higher formability than the ferritic grade. A nitrogen-strengthened version of the austenitic grade, called nitronic, is also available, and several nitronic alloys (e.g., Nitronic 19D and Nitronic 30) can be used for automotive applications. Nitronic 19D is a casting alloy and is recommended for suspension components. Nitronic 30 has excellent formability and is recommended for body panels.

The principal use of stainless steel in today's automobiles is in the exhaust system, where its exceptional corrosion and oxidation resistances give it a considerable edge over steel or aluminized steel. Typical choices for the exhaust system's hot end, including the exhaust manifold, downpipe, and catalytic converter, are the austenitic grades, such as 309 or 310 (25% Cr, 20% Ni). For the cold end, which includes the resonator, intermediate pipe, silencer, and tailpipe, either austenitic grades, such as 304 (18% Cr, 9% Ni), or ferritic grades, such as 409 (12% Cr), are selected.

2.6 Summary

Despite improving composites and lightweight alloys that do not contain iron, steel remains the main material used in producing bodies, transmissions, suspensions, and chassis in serial automobile manufacturing. Such tenacity of manufacturers is explained—metal does not stand still either, progressing from year to year.

Traditional alloys for producing car parts are low-carbon grades, the main structural component of plastic ferrite. The strength limit of such materials does not exceed a modest 365 MPa, which is why they are classified in the automotive industry as a group of so-called mild steels.

High-strength steels or advanced high-strength steels are another direction the global automotive industry is betting on. AHSS are complex materials specially created to meet the most stringent requirements of consumers and automakers for the modern vehicle. This unique lightweight automotive steel is also strong enough to be ideal for producing a safe, environmentally friendly, and inexpensive car of the future. Therefore, according to the experts of WorldAutoSteel, the automotive industry's consumption of AHSS has been growing at the fastest pace in recent years compared to other car materials. For example, we can cite the UltraLight Steel Auto Body (ULSAB) project, whose participants developed the design of an ultralight steel car body design. A consortium of 35 metallurgical companies operating in 18 countries was created to implement the ULSAB initiative, the investment of which amounted to $22 million. They teamed up to offer car manufacturers a new concept of ultra-lightweight steel bodywork. At the same time, the ULSAB project answered the main request of the automotive industry—reducing the weight of a steel car without increasing costs and losing productivity [31].

The share of high-strength, two- and multi-phase, TRIP, and IT steels with innovative coatings is steadily increasing. The release of ultra-low carbon (IF) steels with titanium and niobium alloying improves stamping performance many times, which means that prototypes with unique body shapes and updated layouts are waiting for us soon.

References

1. What is steel? https://worldsteel.org/about-steel/what-is-steel/. Accessed 04 Feb 2025
2. Defining steels. https://ahssinsights.org/metallurgy/defining-steels/. Accessed 04 Feb 2025
3. Horvath CD (2010) 2 – Advanced steels for lightweight automotive structures. In: Mallick PK (ed) Composites science and engineering, materials, design and manufacturing for lightweight vehicles. Woodhead Publishing Series. Woodhead Publishing, UK, pp 35−78. https://doi.org/10.1533/9781845697822.1.35
4. Hovorun TP, Berladir KV, Pererva VI, Rudenko SG, Martynov AI (2017) Modern materials for automotive industry. J Eng Sci 4(2):F8−F18. https://doi.org/10.21272/jes.2017.4(2).f8
5. ULSAB. https://www.worldautosteel.org/projects/ulsab/. Accessed 04 Feb 2025
6. ULSAB and the early steel consortia. https://ahssinsights.org/about/rationale/ulsab-and-the-early-steel-consortia/. Accessed 04 Feb 2025
7. A new global formability diagram. https://ahssinsights.org/blog/a-new-global-formability-diagram/. Accessed 16 Feb 2025
8. Advanced High-Strength Steel (AHSS) definitions. https://www.worldautosteel.org/steel-basics/automotive-advanced-high-strength-steel-ahss-definitions/. Accessed 16 Feb 2025
9. Ebrahimi F, Saeidi N, Raeissi M (2020) Microstructural modifications of dual-phase steels: an overview of recent progress and challenges. Steel Res Int 91:2000178. https://doi.org/10.1002/srin.202000178
10. Martínez-González CJ, López-Baltazar EA, Alvarado-Hernández F, Baltazar-Hernández VH, Saha DC, Biro E, Zhou N (2022) On the measurability and predictability of HAZ softening in GMAW of automotive DP980 steel. Metals 12:1009. https://doi.org/10.3390/met12061009
11. Keeler S, Kimchi M, Mooney PJ (2017) Advanced high-strength steels application guidelines. Version 6.0. WorldAutoSteel
12. You H, Kang M, Yi S, Hyun S, Kim C (2021) Comprehensive analysis of the microstructure and mechanical properties of friction-stir-welded low-carbon high-strength steels with tensile strengths ranging from 590 MPa to 1.5 GPa. Appl Sci 11:5728. https://doi.org/10.3390/app11125728
13. Tisza M (2020) Development of lightweight steels for automotive applications. https://doi.org/10.5772/intechopen.91024
14. Wan C, Ding H, Cai M, Rolfe B (2014) Characterization of microstructures and tensile properties of TRIP-aided steels with different matrix microstructure. Mater Sci Eng A 610:65–75. https://doi.org/10.1016/j.msea.2014.05.029
15. Sakuma Y, Matlock DK, Krauss G (1992) Effects of microstructure and notches on the mechanical properties of dual-phase steels. Metall Trans A 23:1221−1232
16. de Melo Correia Lima R, dos Santos Fernandes Tolomelli FT, Clarke AJ, Clarke KD, Spadotto JC, Rizzo Assunção FC (2022) Microstructural characterization of a 1100 MPa complex-phase steel. J Mater Res Technol 17:184−191. https://doi.org/10.1016/j.jmrt.2021.12.106
17. Zhou W, Hb W, Huo Q et al (2024) Investigation into microstructure and fatigue properties of complex phase steel. J Mater Eng Perform 33:4482–4491. https://doi.org/10.1007/s11665-023-08253-1
18. Li Y, Tang H, Lai R (2021) Microstructure and mechanical performance of resistance spot welded martensitic advanced high strength steel. Processes 9:1021. https://doi.org/10.3390/pr9061021
19. Safa M, Ramadan R, Ibrahim S (2017) Processing and evaluation of ferritic-bainitic multi-phase steel. J Pet Min Eng 19(1):71−80. https://doi.org/10.21608/jpme.2017.39126
20. Cornette D, Cugy A, Hildenbrand A, Bouzekri M, Lovato G (2005) Ultra high strength FeMn TWIP steels for automotive safety parts. Rev Met Paris 102(12):905–918. https://doi.org/10.1051/metal:2005151
21. Yang M, Yue L, Xie K, Zhang S, Sun Y, Tan Y (2020) Study on corrosion behavior of TWIP steel and properties of surface electroless Ni-P coatings. Mater Res Express 7:036517. https://doi.org/10.1088/2053-1591/ab7d0a

22. Mallick PK (2012) 2 – Advanced materials for automotive applications: an overview. In: Rowe J (ed) Advanced materials in automotive engineering. Woodhead Publishing, UK, pp 5–27. https://doi.org/10.1533/9780857095466.5
23. Li J, Tong C, Zhang R, Shi Z, Lin J (2024) A data-informed review of scientific and technological developments and future trends in hot stamping. Int J Lightweight Mater Manuf 7(2):327–343. https://doi.org/10.1016/j.ijlmm.2023.11.003
24. An introduction to the types of hot forming: direct and indirect methods (2024). https://www.weba.at/en/newsroom/blog/forming-technology/types-of-hot-forming-direct-and-indirect-methods-presshardening. Accessed 16 Feb 2025
25. Gramlich A, Schmiedl T, Schönborn S, Melz T, Bleck W (2020) Development of air-hardening martensitic forging steels. Mater Sci Eng A 784:139321. https://doi.org/10.1016/j.msea.2020.139321
26. Zheng K, Lee J, Xiao W, Wang B, Lin J (2018) Experimental investigations of the in-die quenching efficiency and die surface temperature of hot stamping aluminium alloys. Metals 8:231. https://doi.org/10.3390/met8040231
27. Kong HJ, Liu CT (2018) A review on nano-scale precipitation in steels. Technologies 6:36. https://doi.org/10.3390/technologies6010036
28. Carpio M, Calvo J, García O, Pedraza JP, Cabrera JM (2021) Heat treatment design for a qp steel: effect of partitioning temperature. Metals 11:1136. https://doi.org/10.3390/met11071136
29. Mallick PK (2021) Overview (Chap. 1). In: Mallick PK (ed) Materials, materials, design and manufacturing for lightweight vehicles, Second edn. Woodhead Publishing, UK, pp 1–36. https://doi.org/10.1016/B978-0-12-818712-8.00001-X
30. Hariharan K, Balachandran G, Prasad MS (2009) Application of cost-effective stainless steel for automotive components. Mater Manuf Process 24(12):1442–1452. https://doi.org/10.1080/10426910903179989
31. Trends in metallurgy in the automotive industry (2022). https://metinvest-smc.com/ua/articles/tendencii-v-avtomobilestroenii/. Accessed 16 Feb 2025

Chapter 3
Cast Iron

Abstract This chapter explores cast iron's structure, classification, and applications in the automotive industry. Cast iron forms microstructures based on graphite inclusion types, influencing its mechanical properties. Major types include grey, white, ductile, compacted graphite, vermicular graphite, and austempered ductile iron. Grey cast iron is notable for its wear resistance and thermal conductivity, which makes it ideal for engine blocks and brake components. White iron, rich in cementite, is hard and wear-resistant. Ductile and austempered ductile irons offer high strength and ductility and are used in crankshafts, suspension, and chassis parts. Compacted and vermicular graphite irons balance strength and thermal management and are suitable for high-performance engines. Mechanical properties such as tensile strength, ductility, and fatigue resistance vary by composition and treatment. Cast iron remains essential in automotive engineering due to its cost-effectiveness, damping capacity, and machinability, with innovations continuing to expand its application.

3.1 Introduction

The carbon content in cast iron ranges from 2.14% to 6.67%; the rest is iron, impurities, and alloying additives. Carbon can be in graphite or cementite (Fe_3C—cementite, iron carbide) inclusions [1]. Although steel and cast iron contain carbon, the two metals are significantly different. The steel contains less than 2% carbon, which allows the final product to harden into a single microcrystalline structure. The higher carbon content of cast iron means it solidifies as a heterogeneous alloy with more than one microcrystalline structure. Combining high carbon and silicon content gives cast iron excellent casting properties. Different types of cast iron use different methods of heat treatment and finishing.

In technical alloys, the addition of carbon and silicon is expressed as a so-called carbon equivalent ($CE = \%C + 1/3\%Si$), which categorizes cast irons as shown in Fig. 3.1. A CE of 4.3 indicates a eutectic alloy, while alloys with a lower CE are referred to as hypo-eutectic and those with a higher CE are termed hyper-eutectic [2].

K. Berladir et al., *Materials for the Automotive Industry*, Springer Tracts in Mechanical Engineering, https://doi.org/10.1007/978-3-032-05866-9_3

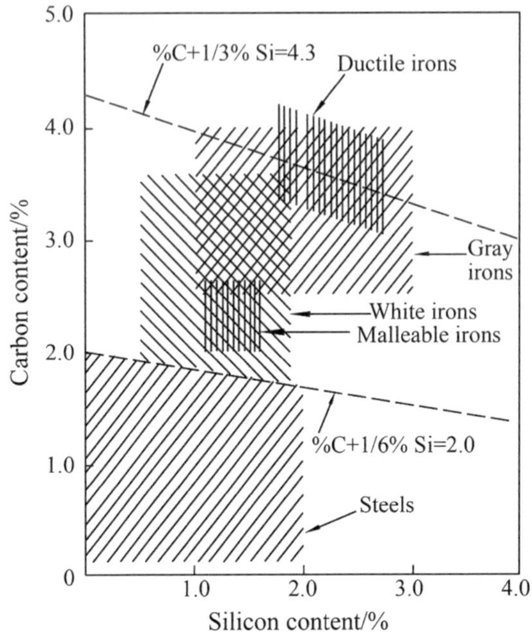

Fig. 3.1 Definition of cast iron alloys according to carbon and silicon contents [2]

The automotive cast iron market includes various types of cast iron used in the automotive industry. Cast iron, the most common, is known for its excellent strength and durability. Grey cast iron has a high carbon content, giving it a grey appearance and increased wear resistance. White cast iron is extremely hard and brittle, making it high-strength and wear-resistant applications. Malleable cast iron is cast iron that has been heat treated to improve it. Other types of iron are on the market, and they have specific properties specially designed for the automotive industry [3].

The main difference between producing malleable and cast iron is that cast iron does not work with hammers and tools. There are also differences in composition—cast iron contains 2–4% carbon and other alloys and 1–3% silicon, improving the molten metal's casting properties. Small amounts of manganese and some impurities such as sulfur and phosphorus may also be present. Differences between malleable iron and cast iron can also be found in the details of the chemical structure and physical properties [4].

Table 3.1 shows the different composition ranges for different types of cast iron.

With the increasing use of high-strength steel and light non-ferrous alloys, the cast iron content of automobiles has decreased significantly over the past few years [6]. Due to its high density compared to steel, cast iron offers no weight-saving advantages. Cast iron is also a material with low ductility. The main advantages of cast iron, which is still used, are its low price, high resistance to wear, damping, and excellent machinability.

In the automobile industry, different types of cast iron parts have emerged to meet the ever-increasing demands of industry. From the classic grey cast iron cylinder

Table 3.1 Range of compositions for typical unalloyed cast irons (%) [1, 5]

Type of iron	C	Si	Mn	S	P
Grey	2.5–4.0	1.0–3.0	0.2–1.0	0.02–0.25	0.02–1.0
Ductile	3.0–4.0	1.8–2.8	0.1–1.0	0.01–0.03	0.01–0.1
Compacted graphite	2.5–4.0	1.0–3.0	0.2–1.0	0.01–0.03	0.01–0.1
Malleable (cast white)	2.0–2.9	0.9–1.9	0.15–1.2	0.02–0.2	0.02–0.2
White	1.8–3.6	0.5–1.9	0.25–0.8	0.06–0.2	0.06–0.2

body to the latest vertical graphite cast iron and isothermally hardened ductile iron, each type plays a critical role in improving performance and pushing the boundaries of automotive technology.

Advantages of cast iron for the automotive industry:

1. Good castability—cast iron has excellent fluidity when melted, allowing it to be cast into complex shapes with intricate details, suitable for casting complex structures or components.
2. High wear resistance—cast iron, especially white cast iron, can provide excellent wear resistance against hard carbides in the structure.
3. Excellent machinability—graphite in cast iron as a natural lubricant reduces wear during operation and facilitates machining. Grey cast iron with graphite flakes provides excellent machinability compared to other types of cast iron.
4. Compressive strength—like cast steel, cast iron also has good strength, making it ideal for buildings and structures that require compressive loads.
5. Low cost—it is generally less expensive than cast steel or cast aluminum, making it a cost-effective option for many applications.

Disadvantages of cast iron for the automotive industry:

1. Brittleness—although cast iron has good strength, it is relatively brittle and can break under impact or pressure, making it unsuitable for applications where toughness is critical.
2. Prone to rust—cast iron rusts very easily due to its high iron content, so its surface is usually painted or plated to protect it from rust.
3. Weight—cast iron is relatively heavy. If weight is an issue, cast aluminum may be an alternative choice.

3.2 Classification of Cast Irons

Depending on the shape of graphite inclusions, three main types of graphite cast iron are distinguished: grey, ductile, and high-strength cast iron (Fig. 3.2).

The presence of graphite inclusions determines the high antifriction and wear-resistant properties of the graphite cast iron. Graphite inclusions act as a solid lubricant and, as it were, lubricate the friction surface, reducing its coefficient of friction

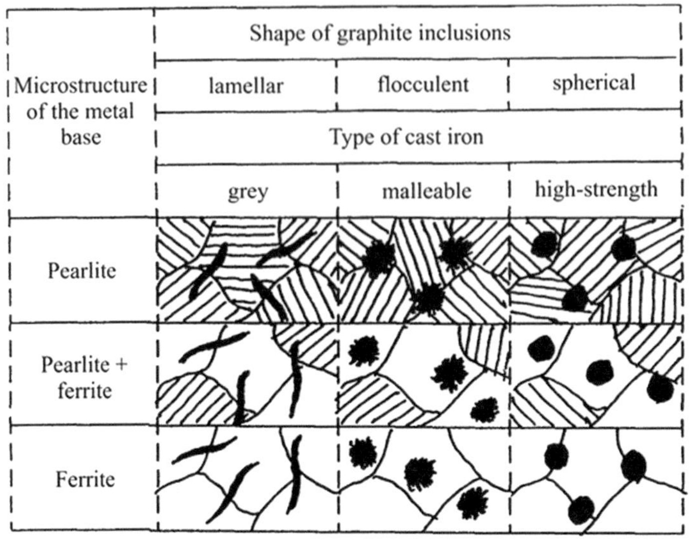

Fig. 3.2 Classification of graphite cast irons according to the shape of graphite inclusions and metal base

[7]. Due to this, the friction surfaces do not stick; they are better worn into each other, and their wear resistance increases. In addition, graphite inclusions increase the strength and absorption (retention ability) of the film of lubricant deposited on the friction surface in lubricated units. Graphite inclusions exposed on the friction surface allow oil to be collected in profile depressions and graphite grains, which helps to increase the resistance of the oil film under high loads and significantly improve its ability to maintain its operating functions [8].

The shape and size of graphite inclusions and the metal base determine the functional properties of cast iron products [9]. Graphite inclusions are peculiar cracks or voids filled with graphite. In this regard, cast iron can be considered steel, riddled with many such cracks and voids (graphite inclusions), weakening the metal base.

The more graphite inclusions, the coarser they are (their shape is more irregular), the more they disconnect the metal base, and the lower the mechanical properties of cast iron. Thus, grey cast iron is brittle and not strong; malleable cast iron has increased ductility compared to others (it takes to shock and vibration loads well), and high-strength cast iron has high-strength characteristics. Also, the mechanical properties of cast iron depend on its metal base, which can be ferritic, ferrite-pearlitic, or pearlitic. Pearlite metal structure is preferable as it has greater strength.

3.2.1 *Grey Cast Iron*

Grey cast iron (GCI), characterized by its high carbon content and characteristic grey fracture due to graphite flakes, has been a staple of the automotive industry for many years. The composition of this metal includes 2.5–4% carbon and 1–3% silicon. The higher hardness compared to other types of grey cast iron is due to the pearlitic matrix [10] (Fig. 3.3). Its unique properties make it a valuable material for various automotive applications [11].

The most famous application of GCI in automobiles is producing engine blocks and cylinder heads. The excellent castability of grey cast iron allows it to be relatively easily shaped into complex shapes. In addition, high thermal conductivity is critical for these parts, as it effectively dissipates the heat generated during engine operation, helping to prevent overheating.

Grey cast iron's excellent wear resistance and high heat capacity make it an ideal material for brake discs and drums. When braking, friction creates a significant amount of heat. The ability of grey cast iron to withstand this heat and dissipate it effectively, as well as its resistance to wear, ensures a long life for these components.

Components such as gears, bearings, and shafts in the transmission system can also be made from GCI. Due to its high wear resistance and excellent machinability, it is suitable for these applications. Suspension and steering parts such as knuckles and control arms are often made of grey cast iron. These parts require a material with good strength, stiffness, and durability to withstand the loads and stresses they are subjected to, and grey cast iron is a perfect fit.

One of the great features of GCI is its ability to absorb vibrations. This is particularly useful in the automotive industry, where minimizing noise and vibration is a key concern. Grey cast iron components help reduce vibration transmission, resulting in quieter, smoother operation.

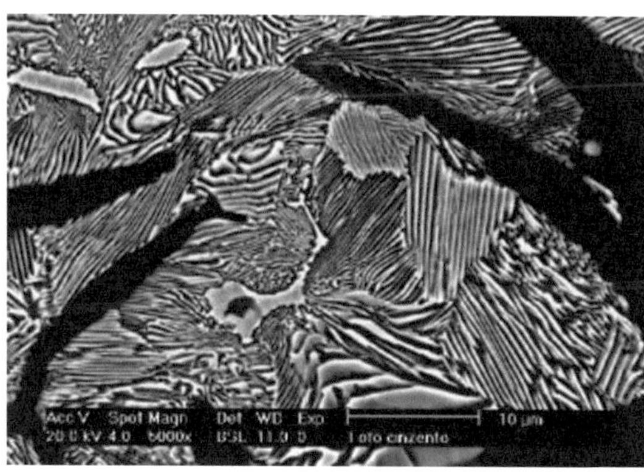

Fig. 3.3 Microstructure of the grey cast iron [10]

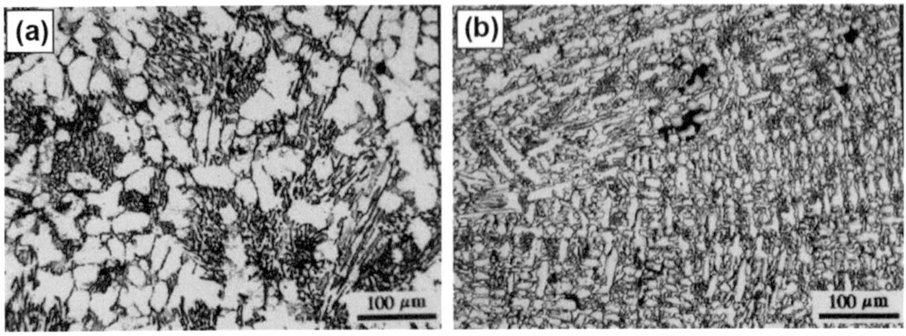

Fig. 3.4 As-cast microstructure of two high chromium white iron alloys: **a** HCWCI_1 (2.66%C, 24.48%Cr), **b** HCWCI_2 (3.29%C, 14.11%Cr) [12]

3.2.2 White Iron

When the carbon content is correct, and the cooling rate is high, the carbon atoms combine with the iron to form iron. The solidified material has little to no free graphite molecules in the hardened material. When cutting white iron, the fracture surface appears white due to the absence of graphite (Fig. 3.4).

The microcrystalline structure of cementite is hard and brittle, with high compressive strength and good wear resistance. In some specialized applications, it is desirable to have white iron on the surface of the product. This can be achieved by using a good heat conductor to form part of the mold. This quickly removes the heat from the molten metal from that particular area while the rest of the casting cools more slowly [13].

One of the most popular brands of white iron is Ni-Hard Iron. Adding chromium and nickel alloys gives this product superior properties for low-impact abrasive sliding applications. White irons and NI-Hard irons fall under a classification of alloys called ASTM A532 the «Standard Specification for Abrasion-Resistant Cast Irons».

White iron can be further processed into malleable iron by heat treatment. The extended heating and cooling program causes the iron carbide molecules to break down, releasing free graphite molecules into the iron. Ductile cast iron with a microcrystalline structure is created by varying the cooling rate and adding alloys [13].

3.2.3 Compacted Graphite Iron

Although aluminum is increasingly used for gasoline engine blocks, grey cast iron still predominates for diesel engine blocks [14]. With the growing trend towards smaller engines and higher cylinder pressures, compact graphite cast iron (CGI) is increasingly used instead of grey cast iron [15]. The graphite particles in CGI have

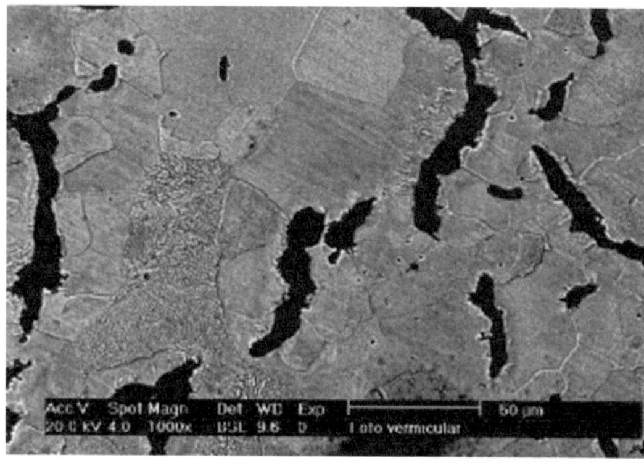

Fig. 3.5 Microstructure of the compacted graphite cast iron [10]

a vermicular or worm-like shape instead of the flaky shape in grey cast iron or the spherical shape in ductile cast iron. Figure 3.5 demonstrates a microstructure with a ferritic matrix, with compacted graphite and a few pearlite grains with very fine pearlite.

As a result, the properties of CGI are between grey cast iron and ductile cast iron. The tensile strength of CGI is 1.5–2 times that of grey cast iron, and the modulus of elasticity of CGI is 150 GPa compared to 105 GPa for grey cast iron. The thermal conductivity of CGI is lower: 38 W/m°K compared to 48 W/m°K for grey cast iron. With higher strength, higher modulus, and lower thermal conductivity, CGI engine blocks can be thinner than grey cast iron engine blocks.

3.2.4 Ductile Iron

Cast irons used in structural automotive applications are ductile iron with a high yield strength (275–625 MPa) and relatively high ductility (2–18% elongation). The modulus of ductile iron is 160–170 GPa, significantly higher than aluminum's.

The ductile cast iron has a ferritic matrix with graphite nodules (Fig. 3.6). Some grains of very fine pearlite can also be seen.

Typical automotive components of ductile iron include the crankshaft, camshaft, center and rear axle covers, various brackets, covers, and other components [14].

The development of pearlitic ductile iron castings (QT700-2, QT800-2, QT800-5) and ferritic ductile iron castings (QT400-18, QT400-15, QT600-10) have been completed. It is obtained primarily by optimizing the smelting process, strictly controlling the content of S, P, and trace elements in the original molten iron, selecting suitable spheroidizing agents, modifiers, alloying elements, and addition methods, and strictly following the modification process in the row.

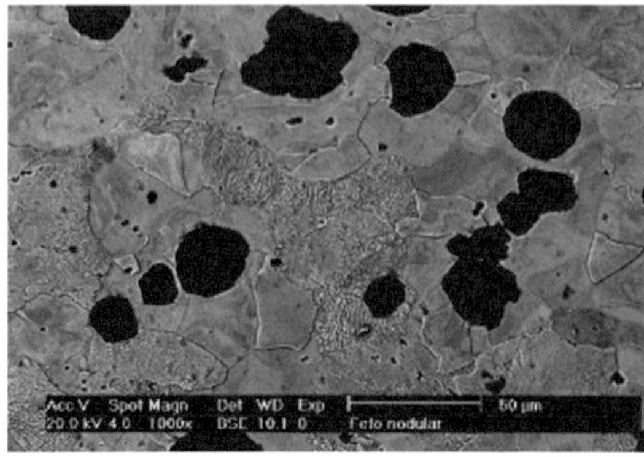

Fig. 3.6 Microstructure of the nodular cast iron [10]

3.2.5 Vermicular Graphite Cast Iron

The microstructures of vermicular iron with different vermicular graphite rates are shown in Fig. 3.7.

A large amount of vermicular graphites and a small amount of globular graphites exist in the structure of vermicular iron. The ratio of the two types of graphite directly affects the comprehensive properties of vermicular iron. As shown in

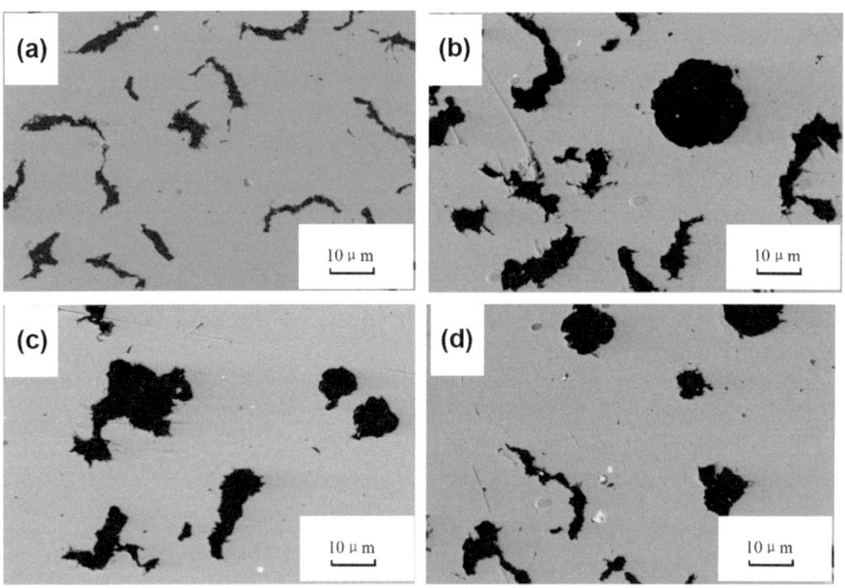

Fig. 3.7 Vermicular iron microstructures with different vermicular graphite rate: **a** 92%, **b** 89%, **c** 74% and **d** 66% [16]

Fig. 3.6, spheroidal graphite appeared in vermicular graphite iron with the decrease of vermicular graphite rate [16].

Vermicular graphite cast iron is widely used in automotive castings such as exhaust pipes and cylinder heads of high-performance engines. The use of cast iron with vermicular graphite for manufacturing cylinder heads of medium and high-power engines allows reducing the weight of the cylinder head due to the design, improving the characteristics of the cylinder head and increasing the engine power. Currently, cast iron with vermicular graphite is expanding in the cylinder blocks of high-performance engines, which makes it possible to reduce the thickness of thin walls, remove reinforcements, and reduce weight by about 25%. It has been successfully used in the cylinder blocks of high-performance engines of foreign companies such as Ford and Audi, and there are also reports of successful use in domestic FAW and Jinan Heavy Duty Truck. The use of pressed cast iron with graphite is limited mainly by the difficulty of controlling its production process in a stable way [5].

SinterCast is a Swedish joint-stock company specializing in research and production equipment based on compacted graphite cast iron technology. Its products (SinterCast Cast Iron Control System) are used on three continents (America, Europe, and Asia) around the world, and it provides technical support in the production of cast iron for 8 of the world's 10 largest car manufacturers. Approximately 20.000 graphite cast iron cylinders or cylinder heads are produced worldwide each month. SinterCast claims that over 95% of the world's cast iron products are produced using its control systems.

3.2.6 Austempered Ductile Cast Iron

The introduction of austempered ductile cast iron (ADI) dates to the mid-1970s in the European commercial vehicle sector. Cast iron has a microstructure containing spherical graphite embedded in a matrix that combines bainitic, ferrite, and austenite and small amounts of martensite and/or carbides [17]. ADI produced by a heat treatment process called aust-tempering has significantly higher yield strength (400–1200 MPa) and fracture toughness than conventional ductile irons. The microstructure of ADI mainly consists of a mixture of carbon-stabilized austenite and acicular ferrite (Fig. 3.8).

This special product phase of austempering heat treatment is called "Ausferrite". The dark needle-like structures are called acicular ferrite and white regions are called carbon-enriched austenite. It is also observed that the microstructure becomes coarse with an increase in austempering temperature [18].

This cast iron is used in the automotive industry for its good machinability, high mechanical properties, wear resistance, and lower processing costs than steel. The yield strength-to-density ratio of ADI is significantly higher than that of cast or forged aluminum. ADI is preferred over aluminum alloys in many chassis and suspension components.

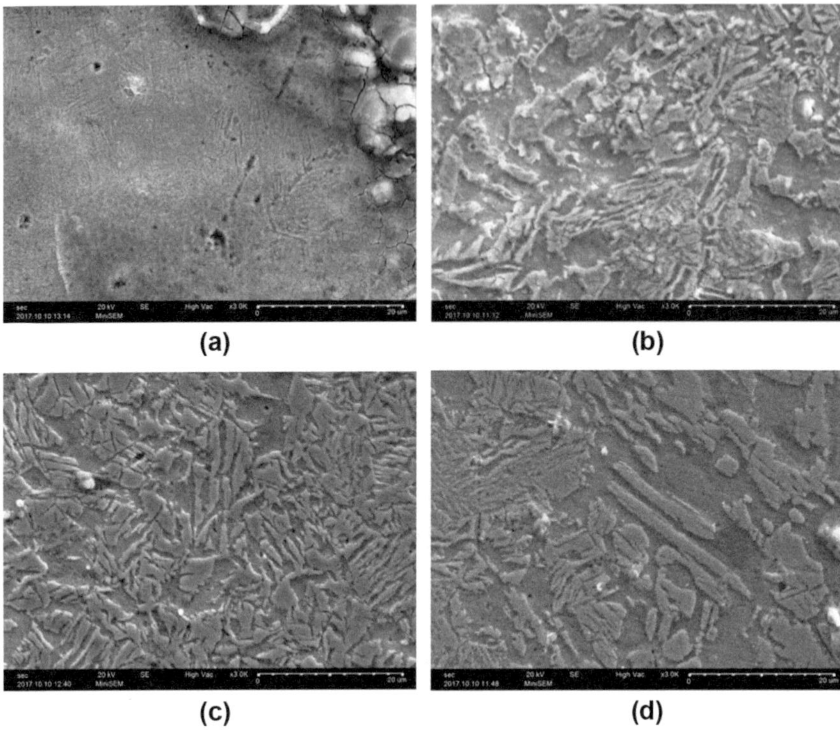

Fig. 3.8 SEM images of ductile cast iron samples austempered at different temperatures: **a** 300 °C; **b** 320 °C; **c** 340 °C; **d** 360 °C [18]

This cast iron typically has a nominal chemical composition of 3.6% C, 2.5% Si, 0.5% Mn, 0.05% Mg, and alloying elements such as Mo, Ni, Cu, Cr, and V. ADI cast iron is mainly used to produce crankshafts, camshafts, gears for various purposes, spring holders, impellers, valve bodies, compressor housings, etc. ADI also exhibits a high damping ability, which causes the parts to absorb 2–5 times more energy than steel parts, reducing noise levels by around 8–10 decibels when using this material in gearboxes.

3.2.7 Development of Special Cast Iron

With the improvement of engine performance, the heat resistance requirements of some castings, such as exhaust pipes, are getting higher, and high silicon molybdenum ductile iron, medium silicon molybdenum ductile iron, and high nickel content ductile iron continue to appear. The FAW Group joint venture mass-produces high-silicon–molybdenum ductile iron exhaust manifolds for the Deutz engine, and some FAW Volkswagen turbo engines have mass-produced high-nickel ductile iron exhaust

Table 3.2 Key mechanical properties for various grades of cast iron

Grades	Brinell hardness	Tensile strength (MPa)	Modulus of elasticity (Msi)	Elongation (%) (in 50 mm)
Grey iron class 25	187	206	16.1	–
Grey iron class 40	235	288	18.2	–
Ductile iron grade 60-40-18	130–170	413	24.5	–
Ductile iron grade 129-90-02	240–300	827	25.5	–
CGI grade 250	Max 179	Min 249		3
CGI grade 450	207–269	Min 449		1

manifolds. Some domestic FAW engines also began using these three cast-iron types [5].

3.3 Mechanical Properties of Cast Irons

A few common mechanical properties of cast iron include:

- Hardness—material's resistance to abrasion and indentation.
- Toughness—material's ability to absorb energy.
- Ductility—material's ability to deform without fracture.
- Elasticity—a material's ability to return to its original dimensions after it has been deformed.
- Malleability—a material's ability to deform under compression without rupturing.
- Tensile strength—the greatest longitudinal stress a material can bear without tearing apart.
- Fatigue strength—the highest stress a material can withstand for several cycles without breaking.

Key mechanical properties for various grades of cast iron are shown in Table 3.2.

3.4 Application

Table 3.3 summarizes typical examples of the application of cast iron in the automotive industry.

Table 3.3 Application of cast iron in the automotive industry [5, 11, 13]

Part of car	Explanation
Brake rotors and calipers	Cast iron is ideal for brake rotors, calipers, and related brackets/mounting components. Automobile brakes generate significant heat and vibration whenever the driver presses the pedal. Cast iron can withstand this intense heat cycle while reducing noise and vibration entering the cabin. The material's improved lubricity and high compressive strength are also ideal for brake performance, ensuring smooth operation and resisting deformation
Suspension and steering components	Vehicle suspension components are also subject to severe vibration and shock when rolling over uneven surfaces. Cast iron suspension and steering components help eliminate noise for a quieter, smoother ride. Examples include suspension brackets, control arms, steering knuckles, and cast rods
Exhaust manifolds	The engine exhaust pipe directs hot gases from the block to the vehicle's rear. Therefore, exhaust pipes or ducts must reliably withstand extreme temperature fluctuations without warping or cracking, which could lead to dangerous exhaust gas leaks. Cast iron is a functional and cost-effective metal for this application, with the added benefit of further reducing exhaust noise and vibration
Engine blocks and cylinder heads	Traditionally, cast iron has been the material for virtually all engine applications. In modern automotive manufacturing, aluminum is often preferred due to its lightness, especially for cylinder heads. However, cast iron is still used extensively for engine blocks, sometimes combined with alloy frames to combine the benefits of cast iron with the weight savings of aluminum
Powertrain and transmission parts	Cast iron is also excellent for powertrain and transmission components, especially housings. The material is durable and economical, easily resists heat from friction, and reduces noise from internal transmission components. In some cases, the transmissions and differentials cast iron, although steel is more common in modern vehicles

3.5 Summary

Cast iron is an iron-carbon alloy with a carbon content between 2.14 and 6.67%. The addition of carbon and silicon influences its microstructure, making it a heterogeneous material. Cast iron is categorized based on its carbon equivalent (CE) into eutectic, hypo-eutectic, and hyper-eutectic alloys. Due to its strength, wear resistance, and machinability, cast iron is widely used in the automotive industry despite its high density and brittleness.

Cast iron is classified based on the shape of graphite inclusions into different types. Grey cast iron (GCI) contains graphite flakes, providing high wear resistance and thermal conductivity, making it suitable for engine blocks, brake discs, and transmission parts. White cast iron is extremely hard and brittle due to the presence of cementite, making it ideal for wear-resistant applications. Compacted graphite iron (CGI) has a vermicular graphite structure, offering strength between grey and ductile iron, and is used in high-performance engine blocks. Ductile iron contains spheroidal graphite, making it stronger and more ductile than grey iron, commonly used in crankshafts, camshafts, and axles. Vermicular graphite cast iron features a mix of vermicular and spheroidal graphite, widely used in high-performance exhaust manifolds and cylinder heads. Austempered ductile cast iron (ADI) is a heat-treated form with enhanced mechanical properties, making it suitable for suspension and chassis components. Special cast irons, such as high-silicon and high-nickel cast irons, have been developed for high-temperature applications like exhaust manifolds.

The mechanical properties of cast iron vary depending on its microstructure and composition. Key properties include hardness, toughness, ductility, tensile strength, fatigue strength, and elasticity. These characteristics make cast iron suitable for a range of industrial applications, particularly in the automotive industry.

In automotive applications, cast iron is commonly used in brake components due to its heat resistance and vibration-dampening properties. It is also used in suspension and steering parts for strength and durability. Exhaust manifolds benefit from cast iron's ability to withstand high temperatures without deformation. Engine blocks and cylinder heads rely on cast iron for its excellent heat dissipation and strength. Additionally, powertrain and transmission components use cast iron to enhance durability and reduce noise.

Despite increasing competition from lighter materials like aluminum, cast iron remains essential for specific automotive components where weight reduction is not the primary concern. Advances in casting technology continue to improve its properties and expand its applications in modern vehicle manufacturing.

References

1. Berns H, Theisen W (2008) Ferrous materials. Steel and cast iron. Springer Berlin, Heidelberg. https://doi.org/10.1007/978-3-540-71848-2
2. Chen XR, Zhai QJ, Dong H et al (2020) Molybdenum alloying in cast iron and steel. Adv Manuf 8:3–14. https://doi.org/10.1007/s40436-019-00282-1
3. The 4 types of cast iron: a comprehensive guide (2025). https://www.shop.machinemfg.com/the-4-types-of-cast-iron-a-comprehensive-guide/. Accessed 08 Feb 2025
4. Cast iron vs malleable iron: what is the main difference? (2024). https://www.plumberstar.com/cast-iron-vs-malleable-iron/. Accessed 08 Feb 2025
5. Fisher R (2024) Introduction to cast iron: history, types, properties, and uses. https://www.reliance-foundry.com/blog/cast-iron. Accessed 08 Feb 2025
6. Czerwinski F (2021) Current trends in automotive lightweighting strategies and materials. Materials 14:6631. https://doi.org/10.3390/ma14216631
7. Seo H, Gweon J, Lee JJ, Sohn SS, Jang H (2024) The effect of solid lubricants in the brake friction composite on brake emissions: a case study with graphite, Sb2S3, and MoS2. Tribol Int 193:109413. https://doi.org/10.1016/j.triboint.2024.109413
8. Morstein CE, Dienwiebel M (2021) Graphite lubrication mechanisms under high mechanical load. Wear 477:203794. https://doi.org/10.1016/j.wear.2021.203794
9. Pokrovskii AI, Chaus AS, Kunovskii EB (2011) Effect of the shape of graphite inclusions on acoustic characteristics of articles from cast and deformed iron. Met Sci Heat Treat 53:311–317. https://doi.org/10.1007/s11041-011-9388-9

10. de Campos MF, Rolim Lopes LC, Magina P, Lee Tavares FC, Kunioshi CT, Goldenstein H (2005) Texture and microtexture studies in different types of cast irons. Mater Sci Eng A 398(1–2):164–170. https://doi.org/10.1016/j.msea.2005.03.018
11. The key role of grey cast iron in automotive manufacturing (2023). https://www.zhycasting.com/the-key-role-of-grey-cast-iron-in-automotive-manufacturing/. Accessed 08 Feb 2025
12. Mihailović M, Patarić A, Jordović B (2023) Heat treatment featuring the key-parameters of high chromium white cast iron microstructure. Metallur Mater Data 1(3):81–83. https://doi.org/10.30544/MMD12
13. Grey cast iron in the automotive industry. https://www.zhycasting.com/grey-cast-iron-in-the-automotive-industry/#:~:text=Engine%20Blocks%3A%20Grey%20cast%20iron,stresses%20experienced%20in%20engine%20operation. Accessed 08 Feb 2025
14. Auto casting guide: 5 types of cast iron parts (2023). https://www.diecasting-mould.com/news/auto-casting-guide-5-types-of-cast-iron-parts. Accessed 08 Feb 2025
15. Automotive iron casting market, global outlook and forecast 2024–2031 (2023). https://www.linkedin.com/pulse/automotive-iron-casting-market-size-growth-forecast-from-2023
16. Guo Q, Zhong Y, Dong T, Gao P, Guo Y, Li J (2018) Effects of vermicular graphite rate on the oxidation resistance and mechanical properties of vermicular graphite iron. J Alloy Compd 765:213–220. https://doi.org/10.1016/j.jallcom.2018.06.236
17. Bai J, Xu H, Wang Y, Chen X, Zhang X, Cao W, Xu Y (2022) Microstructures and mechanical properties of ductile cast iron with different crystallizer inner diameters. Crystals 12:413. https://doi.org/10.3390/cryst12030413
18. Sellamuthu P, Samuel DGH, Dinakaran D, Premkumar VP, Li Z, Seetharaman S (2018) Austempered Ductile Iron (ADI): influence of austempering temperature on microstructure, mechanical and wear properties and energy consumption. Metals 8:53. https://doi.org/10.3390/met8010053

Chapter 4
Aluminum and Its Alloys

Abstract This chapter explores the role of aluminum and its alloys in the automotive industry, highlighting their increasing application in vehicle weight reduction and environmental sustainability. It provides a detailed classification of cast and wrought aluminum alloys, outlining their composition, mechanical properties, and typical automotive uses. The chapter examines historical developments, including the early use of aluminum in engines and body structures, and discusses recent advancements such as Al-Li and Al-Sc alloys. Commonly used alloys are analyzed for their strength, corrosion resistance, and suitability for forming car-body sheets. The chapter also details aluminum's contribution to crash safety, recyclability, and compliance with environmental regulations. Finally, it presents a comprehensive overview of aluminum applications across automotive components, emphasizing its vital role in future lightweight and sustainable vehicle design.

4.1 Introduction

Two options have been used in the last decade to reduce the weight of cars and make them more ecological: high-strength steel or alloyed aluminum alloys [1]. Aluminum alloys have more weight reduction potential than steel, but they are less prone to forming than steel. In addition, products made of aluminum alloys are more expensive than similar products made of steel [2]. Therefore, cars with aluminum bodies are produced only in the premium segment. Recent developments and practical experience show that the use of aluminum alloys can reduce the weight of the base vehicle by 25–40% and the use of high-strength steel by 20–30%.

Aluminum has been used to make cars for over a hundred years (Fig. 4.1). The first aluminum-bodied sports car debuted at the Berlin International Motor Show in 1899. In 1901, Karl Benz, co-founder of Mercedes Benz, designed the first aluminum car engine [3]. Due to the initial difficulties in machining aluminum and its high cost, it took more than 60 years for aluminum to be widely used in the automotive industry. In 1961, the Buick 215 Land Rover was equipped with an aluminum V8 engine. The lightness of the engine was a revelation, and it became an instant hit with racing

K. Berladir et al., *Materials for the Automotive Industry*, Springer Tracts in Mechanical Engineering, https://doi.org/10.1007/978-3-032-05866-9_4

drivers. In 1997, Audi started producing cars with aluminum bodies. Using aluminum made it possible to reduce the body's weight to 239 kg and brought big dividends in reducing fuel consumption.

In 1997, the production of automotive components in Western Europe mainly used cast aluminum alloys (78% share), while wrought aluminum alloys accounted for 22% (Table 4.1).

Fig. 4.1 Timeline of aluminum application in the automotive industry

Table 4.1 The structure of the use of aluminum alloys in the automobile industry of Western Europe

Alloy type	Alloy brand	Mark according to ISO	Share, %
Cast alloys (78%)	A 226	AlSi9Cu3	48
	A 356	AlSi7Mg	20
	A 360	AlSi10Mg	12
	A 413	AlSi12Cu	9
	A 413.2	AlSi12	7
	A 332/A 336	AlSi12CuNiMg	4
Wrought alloys (22%): extrusion, forging, rolling	6060	AlMgSi0,5	35
	6016	AlSi1,2Mn0,4	15
	5754	AlMg3	14
	6082	AlMgSi1	11
	3003	AlMn1	10
	5182	AlMg4,5Mn0,4	9
	7020	AlZn5,4Mg1	6

From the perspective of automotive engineering, aluminum and its alloys have increasingly become prominent lightweight materials used in automobiles in recent years (Fig. 4.2a). In 2006, aluminum surpassed cast iron to become the second most widely used automotive material in the North American market, following steel. Additionally, the market aims for an aluminum content of approximately 570 net pounds per light vehicle by 2030, with cast aluminum alloys expected to contribute over 50% of this total (Fig. 4.2b). Notably, an extruded aluminum bumper beam was emphasized in the crashworthiness design of the Honda Pilot due to its crucial role in impact protection (Fig. 4.2c). To achieve a lightweight structure while maximizing rigidity, Mercedes Benz utilized an intelligent material composition featuring high aluminum content in the bodyshell architecture of the new AMG SL (Fig. 4.2d) [1].

4.2 Classification of Aluminum Alloys

Based in North America, the Aluminum Association (AA) has developed specifications governing the composition, properties, and grading of aluminum alloys [4]. Aluminum alloys are divided into two types: wrought alloys and cast alloys [5].

4.2.1 Cast Aluminum Alloys

Cast aluminum alloys are made by melting pure aluminum and combining it with other metals in liquid form. The mixture is then poured into a sand mold, stamping, or pouring mold. After hardening, the metal is removed from the mold. At this stage, it is in its final form or the form of a billet or ingot for further processing.

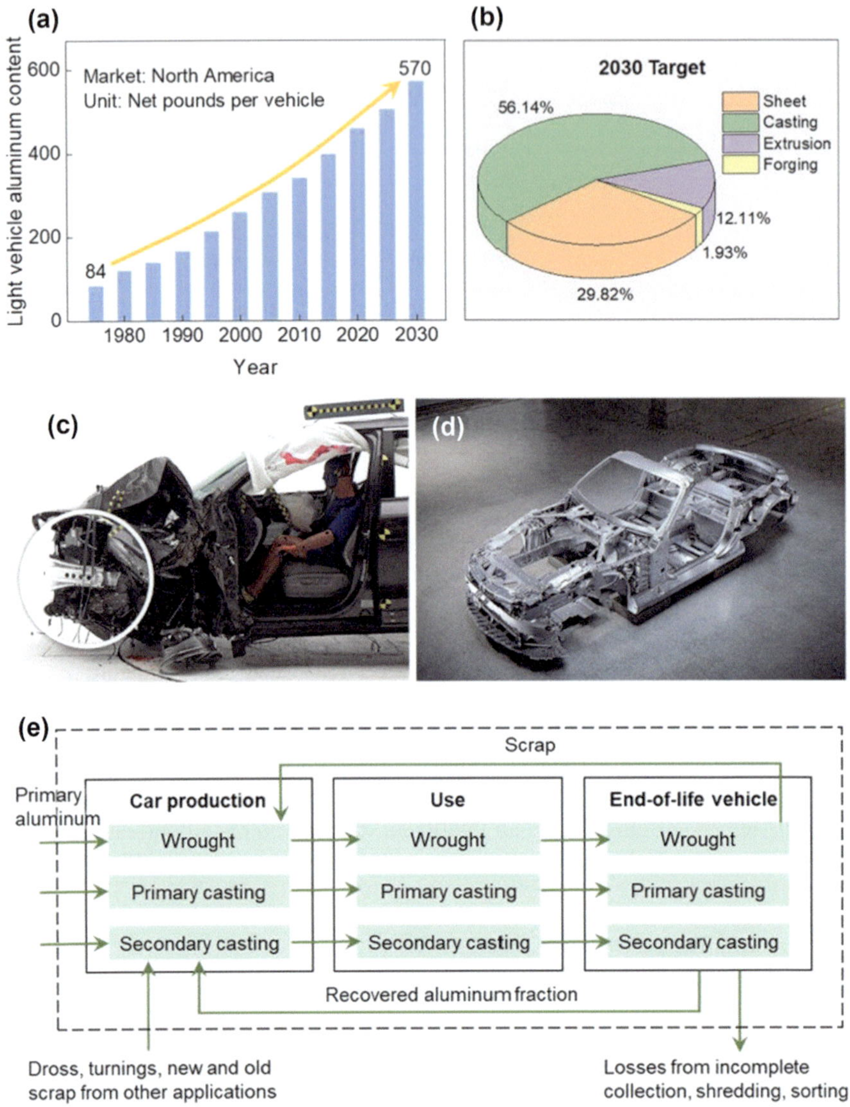

Fig. 4.2 **a** Yearly change of aluminum content in the light vehicle from 1975 to 2030 (North American market), data from light metal age; **b** predicted proportion of different types of aluminum alloy in 2030; **c** Honda pilot frontal crash test (from the Insurance Institute for Highway Safety (IIHS)), where an extruded aluminum bumper beam plays a significant role in impact protection (enlarged area); **d** New Mercedes AMG SL with high aluminum content in the bodyshell architecture to lower weight while achieving maximum rigidity (from Mercedes Benz); **e** illustration of recycling aluminum systems in passenger vehicles [1, 6]

This group of alloys can be divided into heat-treatable and non-heat-treatable aluminum alloys. In addition, these alloys have relatively low tensile strength but allow the creation of economical products. The exceptional casting properties of this type of aluminum alloy are due to the significant presence of silicon in its composition.

The positive properties of aluminum alloys that make them ideal for aluminum alloy casting include:

- Ability to control grain structure.
- Good fluidity.
- Low melting point.
- Ability to strengthen through heat treatment.
- Low solubility of gasses.
- Good surface finish.

At the same time, aluminum alloy has some potentially negative properties. These alloys are susceptible to shrinkage effects due to their high shrinkage characteristics. These alloys, including aluminum castings, also have low ductility and are prone to thermal cracking. They also have a high solubility in hydrogen gas.

Alloys are classified and named depending on the amount and type of alloying metals. The AA naming convention for these alloys consists of four digits, with the third and fourth separated by a decimal point, as follows (XXX.X). The first number indicates the main alloying metals for each aluminum alloy and is the most important. You can use it to distinguish between alloys, «series» or «grades» (Table 4.2).

For example, all aluminum alloys that start with a 2 are known as 2xx.x series or grade 2xx.x alloys. This digit can take any number from 1 to 8, except for 6, which is not used for cast alloys.

The second and third digits in the AA naming convention are assigned arbitrarily to identify the different alloys in each series. The exception is the 1xx.x series, which indicates the decimal percentage of the aluminum content. For example, 160.0 represents a casting containing at least 99.60% aluminum. The fourth digit after the decimal point indicates whether the alloy is cast (xxx.0) or ingot (xxx.1).

4.2.2 Wrought Aluminum Alloys

Wrought aluminum alloys also begin by combining molten aluminum with other metals. However, cast alloys are formed into their final shape by extrusion, rolling, and bending after the metal has solidified into billets or ingots [7].

They are heat-treatable and non-heat-treatable. Wrought aluminum alloys make up about 85% of all aluminum alloys. Alloying elements such as silicon, copper, zinc, and magnesium provide the initial strength of the heat-treatable alloy. They can also be strengthened by heat treatment. Non-heat-treatable forms have an initial strength of alloying elements such as silicon, manganese, and magnesium. These alloys can be further strengthened by quenching or cold working [8].

Table 4.2 Designation system of cast aluminum alloys (by AA) [5]

Series	Main alloying element(s)	Main characteristics	Example application
1xx.x	99.xx% pure aluminum	High electrical and thermal conductivity, excellent corrosion resistance	Big electrical rotors
2xx.x	Copper and small quantities of magnesium	High strength but low corrosion resistance	Cylinder heads and pistons, housings, and bearings
3xx.x	Silicon, copper, and/or magnesium	High strength and wear resistance. Good corrosion resistance	Motor parts, structural parts, marine and aircraft castings
4xx.x	Silicon	Moderate strength, high ductility, and good impact resistance	Bridge railing support castings, dental equipment, cookware
5xx.x	Magnesium	Moderate-to-high strength, high corrosion resistance, good machinability, and attractive appearance	Architectural and ornamental castings, welded assemblies
7xx.x	Zinc	Good finish, good corrosion resistance, high strength through heat treatment	Automotive parts, mining equipment, castings for brazing
8xx.x	Tin	Low friction	Bearing and bushing applications

Their classification and naming system is similar to cast alloys. They are classified into series depending on the main alloying elements. Their names use four digits (but no decimal point), with the first digit indicating the main alloying elements (Table 4.3).

The second digit in the wrought alloy rating, if not 0, tells you if the alloy is a variation of the base alloy, i.e. j. alloy 6160 is the first variation of alloy 6060. The third and fourth digits are assigned randomly to identify the alloys in their respective series, again except for the 1xxx series alloys, where the last two digits describe the purity of aluminum in the alloy.

There are many minor differences between wrought and cast aluminum alloys [7]. For example, cast alloys may contain larger amounts of other metals than wrought alloys. However, the most significant difference between these alloys is the manufacturing process by which they obtain the final product.

Apart from some surface treatment, cast alloys leave the mold in almost the desired solid form, while wrought alloys undergo few modifications in the solid state. This difference will have a significant impact on the possible shapes and physical properties of the final products.

Table 4.3 Designation system of wrought aluminum alloys (by AA) [5]

Series	Main alloying element(s)	Main characteristics	Example application
1xxx	99.xx% pure aluminum	High electrical and thermal conductivity, excellent corrosion resistance	Electrical conductors and chemical processing equipment
2xxx	Copper	High strength-to-weight ratio, low corrosion resistance	Truck wheels and suspensions, aircraft fuselage, and wings
3xxx	Manganese	Moderate strength and good workability	General sheet work, recreation vehicles, electronics
4xxx	Silicon	Low melting point and thermal expansion, high wear resistance	Welding wire and brazing alloy, architectural applications, forged engine pistons
5xxx	Magnesium	Moderate-to-high strength, good weldability, good corrosion resistance	Appliances, automotive parts, marine components
6xxx	Silicon and magnesium	Medium strength with good formability, weldability, machinability, and corrosion resistance	Structural applications, architectural extrusions, recreational equipment
7xxx	Zinc	Moderate-to-very high strength	Airframe structures, mobile equipment, high-stress parts

4.2.3 *Perspective Aluminum Alloys*

At the beginning of the twentieth century, the automotive industry widely used deformable aluminum alloys AA5754 H22, AA6082 T6, and AA7075. Alloy AA5754 (AlMg3) contains 2.60–3.60% Mg, 0.10–0.60% (Mn + Cr), not more than 0.20% Zn and not more than 0.15% Ti.

This alloy has an average strength among aluminum alloys: 220–270 MPa and is sold in sheet and plate form. Alloy 5754, manufactured by Aalco Metals Ltd, UK, is highly corrosion-resistant in seawater and industrial atmospheres. It is used in sports cars like the Jaguar XK, Lotus Evora, and Chevrolet Corvette.

At the end of the twentieth century, the casting aluminum alloy AlSi9Cu3 became the most widespread in the automotive industry, the share of which was almost 50% of all castings. AlSi7Mg (20%) and AlSi10Mg (12%) alloys are in second and third place. A

Alloy AA6082 T6 (AlSi1MgMn) contains 0.70–1.30% Si, 0.60–1.20% Mg, 0.40–1.00% Mn, not more than 0.25% Cr, not more than 0.20% Zn, maximum 0.10% Ti. Available in sheets and plates. It has a tensile strength of at least 275–300 MPa and welds well. It is widely used to produce body structural parts by extrusion. Such parts are used in luxury cars such as the BMW 6, Jaguar XJ, Jaguar XK, Range Rover, and Rolls-Royce Phantom.

Alloy AA7075 (AlZn5.5MgCu) belongs to the deformable high-strength aluminum alloys used in transport (automobiles, aviation, etc.). Its typical chemical composition is as follows: 5.1–6.1% Zn, 2.1–2.9% Mg, 1.2–2.0% Cu, 0.18–0.28% Cr, not more than 0.4% Si, not more than 0.2% Ti, not more than 0.3% Mn. The tensile strength of the alloy is 346–540 MPa. The disadvantage of the alloy is low corrosion resistance and malleability at room temperature.

Alloys of the aluminum–lithium and aluminum-scandium systems are promising for the automotive industry. Thus, the third generation of aluminum–lithium (Al-Li) alloys can manufacture luxury car components due to their low density, high strength, and high stiffness. It should be noted that in the past, these alloys were reserved for use in the aerospace industry. Examples of Al-Li alloys are the high-strength alloys 2195 (Ai-Cu-Li-Mg) and 2050 (Al-Cu-Li), developed specifically to meet the needs of the aerospace and defense sectors [9].

Aluminum alloy with scandium (Al-Sc) is characterized by high strength, high corrosion resistance, and good weldability. In addition, it retains its characteristics at high temperatures, especially in combination with zirconium. These properties of the alloy can be achieved by adding scandium in the range of 0.1–1.0 wt. %. These alloys are quite expensive and more focused on applications in the aerospace industry. However, the further development and reduction of scandium production costs increase the likelihood of their use in the automotive industry.

4.3 Aluminum Alloys for Car Body Sheets

The low density, good machinability, strength, and durability of aluminum alloys meet the requirements for materials for car bodies and other road equipment. These properties also meet increasingly stringent environmental regulations regarding pollution and easy and cheap recycling.

Aluminum in cars mainly reduces weight, reducing environmental problems related to fuel consumption. A car that is ten percent lighter will save 5% on fuel. A weight reduction of just 1 kg reduces CO_2 emissions by 20 kg over 170,000 km driven.

Some chemical compositions and properties of alloys are given in Tables 4.4 and 4.5 [10–12].

Replacing the steel sheet with aluminum means reducing the weight of the car body by 40–50%. The lower rigidity of aluminum sheets is one of the rare disadvantages compared to steel, but this can be compensated for by increasing the thickness of the sheets by up to 50%, which also increases the safety of cars. The strength indicators of aluminum materials for car bodies are lower than those of deep-drawn steel sheets. Due to the greater thickness of the aluminum sheets to achieve rigidity and the aging that can occur during painting, the obtained strength values are comparable to the strength values of steel.

The production of aluminum alloy sheets for car bodies is more expensive than steel sheets due to lower deep drawing capacity, more work steps, and higher quality

Table 4.4 Chemical compositions of aluminum alloys for car body sheets (mass %)

Group	Designation	Si	Cu	Mn	Mg
AlMg	AA6111	0.4	0.1	0.5	2.6–3.6
AlMg	AA5182	0.2	0.15	0.2–0.5	4.0–5.0
AlMgSi	AA6016	1.0–1.5	0.2	0.2	0.25–0.6
AlMgSi (Cu)	AA6009	0.6–1.0	0.15–0.6	0.2–0.8	0.4–0.8
AlMgSi (Cu)	AA6111	0.7–1.1	0.5–0.9	0.15–0.45	0.5–1.0
AlMgSi (Si)	AA2036	0.5	2.2–3.0	0.1-–0.4	0.3–0.6

Table 4.5 Mechanical properties of some alloys for car-body sheets

Alloy	Tempering	R_m, MPa	$R_{p0.2}$, MPa	A_5, %	n	R
AA5754	Annealed	210	100	28	0.30	0.75
AA5182	Annealed	280	140	30	0.31	0.75
AA6009	T4	230	125	27	0.23	0.70
AA6016	T4	240	120	28	0.27	0.65
AA6111	T4	275	160	28	0.26	0.56
AA2036	T4	340	195	24		

tooling required. From the point of view of weight reduction, energy balance, and recycling, the use of aluminum is already a great advantage in the life cycle of cars.

Different manufacturers have developed different alloys for car body sheets. Their origin is based on standard wrought alloys. Alloys of the AlMg group (AA5xxx) do not undergo aging and are classified as materials with average strength values. Machinability and strength indicators depend on the magnesium content: increased magnesium content increases strength and reduces machinability. Soft hardened sheets are more suitable for forming and deep drawing than AlMgSi alloys. The most commonly used alloys are AlMg2.5, AlMg3 and AlMg5Mn. Some alloys of group AA5xxx also contain copper. Copper increases strength due to aging. At the same time, copper reduces its corrosion resistance due to the precipitation of Al2CuMg.

Alcoa developed a frame space concept for producing car bodies based on hollow profiles. Profiles are made of AlMgSi alloys (AlMgSi0.5) and are produced with an extrusion ratio of 20:1. The sections are joined into units made of AlSi alloys (AlSi10Mg) by injection molding. The frame weighs 130–150 kg.

Aluminum car bodies produced in this way comprise 20% castings, 25% pressings, and 55% sheets. Further development of aluminum materials for car bodies will focus on improving machinability and increasing strength properties, with yield strength being the priority. Further development aims to reduce the size of the extruded parts to reduce weight and cost further.

4.4 Application

Cars are complex machines that have upwards of 25,000 parts, and many of them can be made of aluminum [10]. Figure 4.3 presents examples of the potential of 6000 series aluminum alloys used in new energy vehicles.

Table 4.6 summarizes the typical application of aluminum alloys for car production.

Main alloys are 1100, 2024, 3003, 3004, 3105, 4032, 5005, 5052, 5083, 5182, 5251, 5754, 6016, 6061, 6181, 7003, and 7046. Aluminum is available in various formats and tempers: sheets, plates, bars, and tubes [14, 15].

4.5 Summary

The increasing use of aluminum and its alloys in the automotive industry has significantly reduced vehicle weight, improved fuel efficiency, and enhanced sustainability. While the initial adoption of aluminum was slow due to machining difficulties and costs, technological advancements have enabled its widespread application, particularly in premium vehicle segments. The transition from cast to wrought aluminum alloys and the development of high-strength and corrosion-resistant compositions have further expanded their utility in manufacturing lightweight yet durable car components.

Classification systems such as those established by the Aluminum Association have facilitated the identification and selection of suitable aluminum alloys for different automotive applications. The distinction between cast and wrought alloys, each with unique mechanical properties and forming techniques, has allowed manufacturers to optimize material usage for specific vehicle parts. Additionally, innovations in aluminum–lithium and aluminum-scandium alloys demonstrate the potential for further advancements, particularly in electric and high-performance vehicles, where strength-to-weight ratios and durability are critical.

As the automotive industry moves toward stricter environmental regulations and electrification, aluminum alloys will continue to play a vital role in vehicle design and production. Their recyclability, lightweight nature, and adaptability to evolving manufacturing processes make them indispensable in the pursuit of more sustainable transportation. While challenges remain in terms of cost and formability compared to traditional steel, ongoing research and alloy development are expected to enhance their feasibility, ensuring aluminum remains a cornerstone material in future automotive engineering.

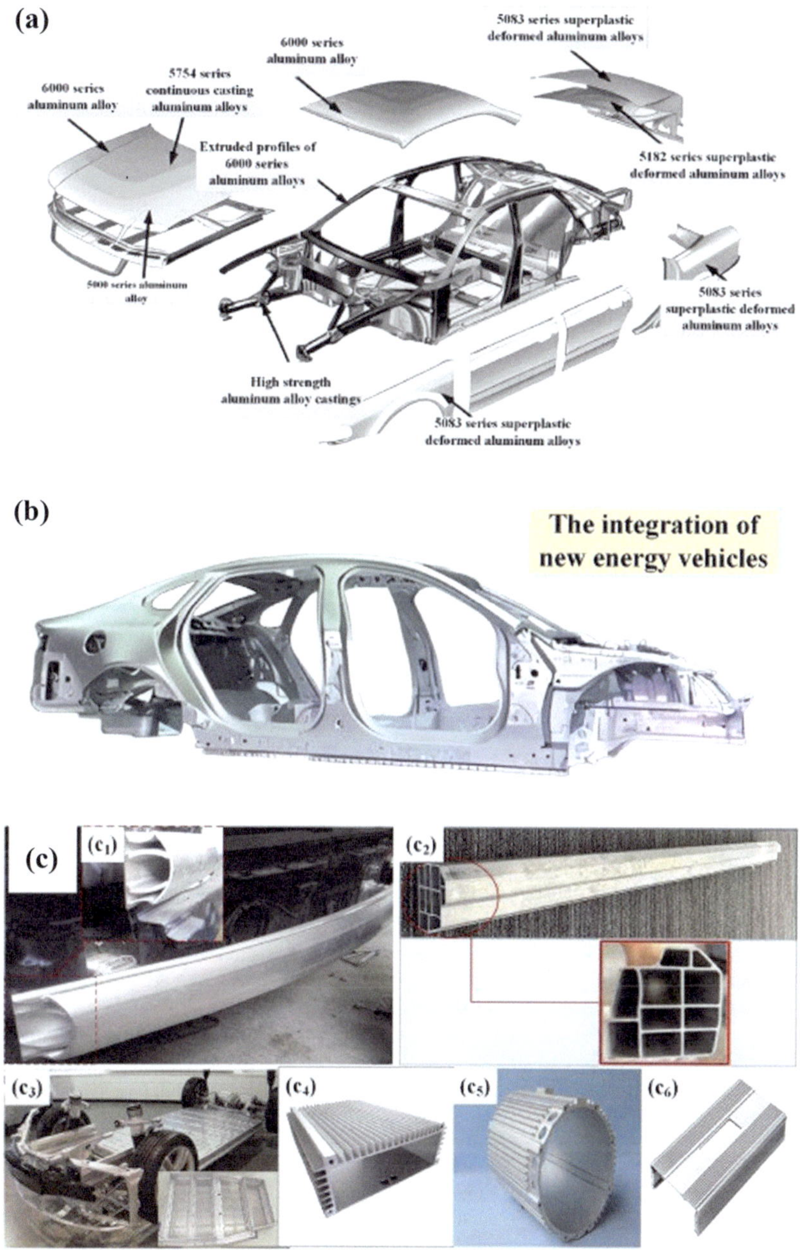

Fig. 4.3 **a** Aluminum alloys for new energy vehicle applications; **b** integration of new energy vehicles; **c** application of 6000 series aluminum alloy profiles or plates: $\mathbf{c_1}$ bumper beam, $\mathbf{c_2}$ door sill beams, $\mathbf{c_3}$ battery tray, $\mathbf{c_4}$ battery pack casing, $\mathbf{c_5}$ motor housing, $\mathbf{c_6}$ automobile cooling plate [13]

Table 4.6 Application of aluminum in the automotive industry

Car part	Type of aluminum	Application
Auto body	3105, 5005, 5083, 5182, 6016, 6181	Body components: hood, quarter panel, fender, bumper, spoiler, roof rack, grille, cowling, decklid, header panel, pillars, rims and hubcaps, trunk, valance
		Doors: central locking, hinges, latches, handles
		Windows: motors, sunroof, sunroof railing
Electronics	1350	Audio and video devices, backup cameras and displays, electrical supply system, gauges and meters, ignition system, lighting and signaling, sensors, switches, wiring harnesses
Interior	1100, 3105, 5052	Floor components and parts, car seats, heat insulators
Powertrain and chassis	2024, 4032, 5052, 5005, 5083, 6016, 6082, 6111	The braking system, electrified powertrain components, engine components, engine oil system, engine cooling system, fuel supply system, suspension and steering, transmission system
Miscellaneous auto parts	1100, 3105, 5052	Air conditioning, bearings, hose

References

1. Zhang W, Xu J (2022) Advanced lightweight materials for automobiles: a review. Mater Des 221:110994. https://doi.org/10.1016/j.matdes.2022.110994
2. Tisza M, Czinege I (2018) Comparative study of the application of steels and aluminium in lightweight production of automotive parts. Int J Lightweight Mater Manuf 1(4):229–238. https://doi.org/10.1016/j.ijlmm.2018.09.001
3. History of Automotive Aluminum (2024). https://hdmetalm.com/blog/history-of-automotive-aluminum/. Accessed 17 Feb 2025
4. Industry Standards. https://www.aluminum.org/industry-standards. Accessed 17 Feb 2025
5. Aluminum Grades: What are the types of aluminum, and how are they classified? https://www.gabrian.com/different-types-of-aluminum-grades/. Accessed 17 Feb 2025
6. Modaresi R, Müller DB (2012) The role of automobiles for the future of aluminum recycling. Environ Sci Technol 46(16):8587–8594. https://doi.org/10.1021/es300648w
7. Differences between cast and wrought aluminum alloys. https://www.howardprecision.com/differences-between-cast-and-wrought-aluminum-alloys/. Accessed 17 Feb 2025
8. Čuha D, Hatala M (2022) Effect of a modified impact angle of an ultrasonically generated pulsating water jet on aluminum alloy erosion using upward and downward stair trajectory. Wear 500–501:204369. https://doi.org/10.1016/j.wear.2022.204369
9. Lakshmi A, Buddi T, Subbiah R, Bandhavi C (2020) Formability studies of automotive aluminium alloy sheet series: a review. E3S Web Conf 184:01036. https://doi.org/10.1051/e3sconf/202018401036

10. Long RS, Boettcher E, Crawford D (2017) Current and future uses of aluminum in the automotive industry. JOM 69:2635–2639. https://doi.org/10.1007/s11837-017-2554-9
11. Miller W, Zhuang L, Bottema J, Wittebrood A, Smet P, Haszler A, Vieregge A (2000) Materials science and engineering. Recent development in aluminum alloys for the automotive industry. Mater Sci Eng A 280(1):37–49. https://doi.org/10.1016/S0921-5093(99)00653-X
12. Montijo S (2021) What is aluminum used for: automotive edition. Kloecknermetals. https://www.kloecknermetals.com/blog/what-is-aluminum-used-for-automotive-edition/. Accessed 17 Feb 2025
13. Gao Y-C, Dong B-X, Yang H-Y et al (2024) Research progress, application and development of high performance 6000 series aluminum alloys for new energy vehicles. J Mark Res 32:1868–1900. https://doi.org/10.1016/j.jmrt.2024.08.018
14. Musfirah A, Jaharah A (2012) Magnesium and aluminum alloys in automotive industry. J Appl Sci Res 8(9):4865–4875
15. Hirsch J (1997) Aluminium alloys for automotive application. Mater Sci Forum 242:33–50. https://doi.org/10.4028/www.scientific.net/msf.242.33

Chapter 5
Magnesium and Its Alloys

Abstract This chapter explores the properties, development, and automotive applications of magnesium and its alloys. Magnesium offers superior castability, high specific strength, and exceptional vibration damping as the lightest structural metal. Alloying elements such as aluminum, zinc, manganese, and rare earths enhance its mechanical properties and corrosion resistance. Historical applications in the automotive and aerospace industries have evolved. Magnesium components are now featured in engine blocks, transmission cases, wheels, seat frames, and structural body parts, significantly reducing vehicle weight. Despite challenges like flammability and creep resistance, continuous alloy development and advanced manufacturing have driven magnesium's resurgence. Detailed industrial examples illustrate its integration in body, chassis, powertrain, and interior systems. The chapter underscores magnesium's critical role in lightweight, fuel-efficient vehicle design and the ongoing evolution of high-performance magnesium alloys.

5.1 Introduction

Magnesium is the lightest of all technical metals; its density is 1.74 g/cm^3. It is 35% lighter than aluminum (2.7 g/cm^3) and more than four times lighter than steel (7.86 g/cm^3) [1]. Magnesium is the eighth most abundant element. It has good ductility, better noise and vibration absorption characteristics than aluminum, and excellent castability. Alloying magnesium with aluminum, manganese, rare earth, thorium, zinc, or zirconium increases the strength-to-weight ratio, making them important materials for applications where weight reduction is important, and reduction of inertial forces is critical [2]. Thanks to this property, denser materials, not only steel, cast iron, and copper-based alloys, but even aluminum alloys, are being replaced by magnesium-based alloys. The requirement to reduce the weight of vehicle components due to legislation to limit emissions has renewed interest in magnesium [3].

The largest share of consumption of magnesium alloys is in the automotive industry—around 70% [4]. Magnesium alloy automotive parts are diverse: structural parts (seat frame and front frame), heat-resistant parts (cylinder block), and

K. Berladir et al., *Materials for the Automotive Industry*, Springer Tracts in Mechanical Engineering, https://doi.org/10.1007/978-3-032-05866-9_5

moving parts (wheels), all of which have the common characteristic of bearing low mechanical and chemical loads.

The earliest use of magnesium and its alloys in automotive engineering dates to 1918, when magnesium-based racing engine pistons were introduced for the Indy 500. However, it was not until the 1930s that Volkswagen fitted a magnesium-based engine block to its Beetle series, marking the beginning of a wide use of magnesium in the automotive industry [5]. Since then, magnesium has been used in various powertrains, chassis, and body structures. Magnesium remains the third most used metal material in the automotive industry.

The reduced weight of the structure and the high specific strength of Mg contributed to its use in racing cars. Volkswagen was the first to use Mg to transmit its Beetle model in 1936, and production peaked in 1971. Porsche also developed Mg casting technology for engine blocks in 1968. Historical records of Mg use in bombers (the B-36 Bomber contained 19,000 pounds of magnesium) and rockets (the Titan I rocket used 1100 pounds of magnesium sheet). Over the years, magnesium alloys have expanded in civil and military aviation [6].

Figure 5.1 summarizes the historical development of automotive magnesium alloy components and provides practical examples from original equipment manufacturers. The development of Mg alloy is closely related to politics, environment, demand, and other factors.

Although using Mg provided good mileage in the automotive and aerospace industries, there were significant problems with high-temperature creep, flammability, room-temperature ductility, and corrosion. As a result, in the 1980s, scientists focused on the development of magnesium alloys, which led to an increase in the use of magnesium in the automotive industry.

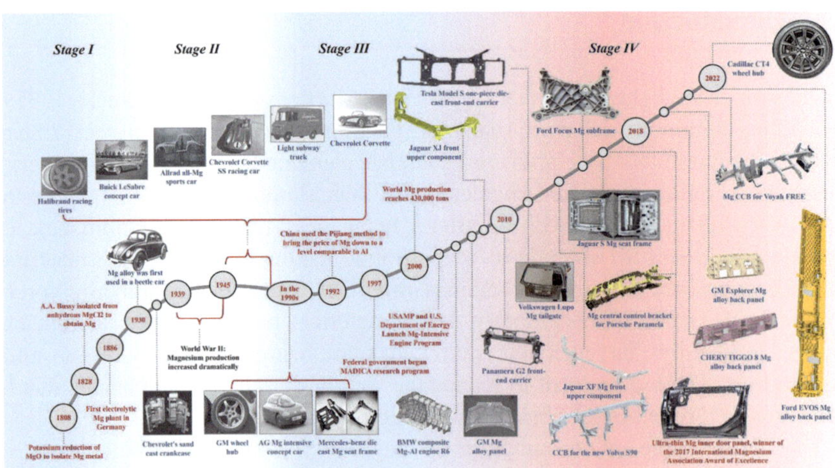

Fig. 5.1 Timeline of magnesium application in the automotive industry [4]

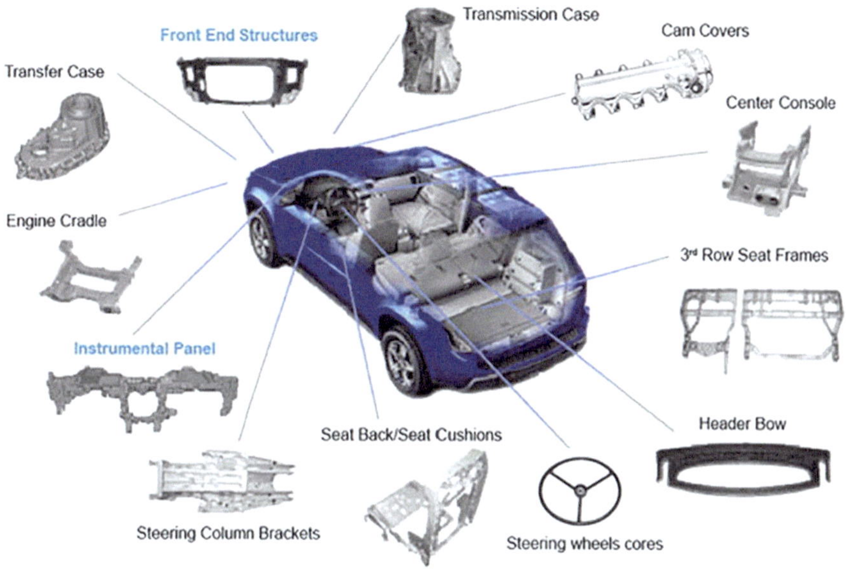

Fig. 5.2 Use of magnesium-based materials in the automotive industry [5]

The use of magnesium alloys in automobiles can be attributed to engine and transmission housings, interior parts, chassis, and body parts (Fig. 5.2).

5.2 Magnesium Alloys

Magnesium is alloyed with other metals for all structural applications to provide adequate strength, corrosion resistance, formability, etc. Typical alloying elements and impurities and their corresponding effects on magnesium are as follows [7–9]:

- Aluminum is the most common alloying element. It increases strength, hardness, and corrosion resistance but reduces ductility. An aluminum content of around 5–6% provides the optimum combination of strength and ductility for structural applications. Increasing the aluminum content extends the solidification range, making the alloy easier to cast but making extrusion more difficult due to increased hardness.
- Zinc is used to increase strength. However, it reduces ductility, increases hot embrittlement of Mg-Al alloys, and can lead to microporosity and accelerate corrosion.
- Manganese slightly increases the yield strength but does not affect the tensile strength. Its most important function is to improve the corrosion resistance of Mg-Al alloys by removing iron and other heavy metal elements into relatively harmless intermetallic compounds, some released during melting.

- Rare earth additives strengthen precipitation and can improve strength and ductility by changing the crystallographic texture. Rare earth elements can also improve creep resistance.
- Silicon increases fluidity and slightly improves creep strength.
- Zirconium is an important grain refiner for sand and metal molds, gravity casting, and low-pressure casting of magnesium alloys.
- Tin increases ductility.
- Lithium reduces density and can improve ductility by forming a cubic structure at additions greater than 11%. The addition of lithium significantly increases cost and reduces corrosion resistance.
- Thorium and yttrium increase creep resistance. Alkaline earth calcium and strontium have also been shown to improve creep resistance.
- Iron reduces the corrosion resistance of magnesium alloys and is kept below 50 ppm in all alloys used today (ASTM International, 2008).
- Similarly, copper and nickel reduce the corrosion resistance of magnesium alloys. Copper is lettered due to its use in ZC63, an early sand-casting alloy.

Magnesium alloys are designated by a combination of letters and numbers that describe the main alloying elements and the percentage of these elements. ASTM B 951 specifies a protocol for using letters and numbers to identify magnesium alloys. According to this standard, groups of magnesium alloys are designated by two letters indicating the main alloying elements (Table 5.1).

Alloy letters indicate the two alloying elements with the highest concentration and are arranged in descending order of composition. Following these letters are numbers indicating the amount of each of the two alloying elements.

The most famous magnesium alloy is AZ91. This magnesium alloy contains 9% (by wt.) Al and 1% Zn. It also contains Mn, but it is not mentioned in the name. Other examples include AZ31, which is 3% Al and 1% Zn, and ZK30, which is 3% Zn and less than 1% Zr (Table 5.2). After the numbers, a letter may appear indicating the rank of the alloy in the chronology of the development of the given alloy. Therefore, AZ91B is a newer version of AZ91A, etc.

Figure 5.3 illustrates the microstructure of the cast magnesium alloys (a) AZ31 and (b) AZ91. The AZ31 alloy displays a globular structure, whereas the AZ91 alloy reveals a fully developed dendritic structure. This difference can primarily

Table 5.1 Letters represent alloying elements for magnesium [10]

A—aluminum	J—strontium	R—chromium
B—bismuth	K—zirconium	S—silicon
C—copper	L—lithium	T—tin
D—cadmium	M—manganese	V—gadolinium
E—rare earths	N—nickel	W—yttrium
F—iron	P—lead	Y—antimony
H—thorium	Q—silver	Z—zinc

Table 5.2 Chemical composition of the AZ31 and AZ91 compared to the standard in wt.% [11]

Alloy	Al	Zn	Mn	Mg
AZ31	2.8	0.7	0.3	96.2
AZ31 standard (DIN EN 12,438)	2.5–3.5	0.6–1.4	0.2–1.0	–
AZ91	9.6	0.6	0.2	89.6
AZ91 standard (ASTM B93)	8.5–9.5	0.45–0.9	0.17–0.4	–

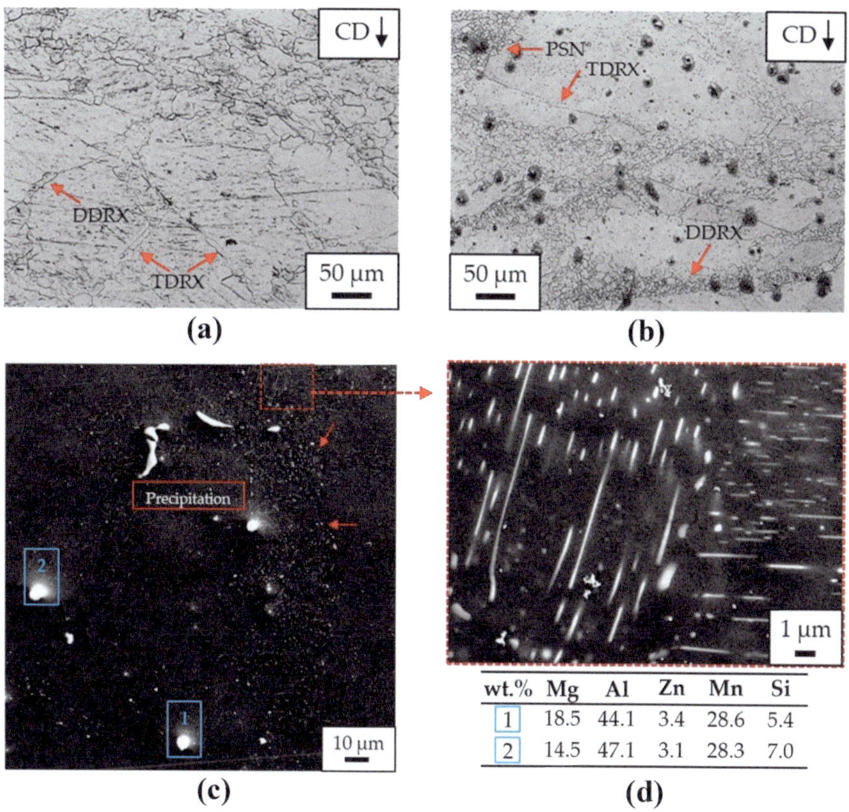

wt.%	Mg	Al	Zn	Mn	Si
1	18.5	44.1	3.4	28.6	5.4
2	14.5	47.1	3.1	28.3	7.0

Fig. 5.3 Microstructure of the hot-deformed magnesium alloy **a** AZ31 and **b** AZ91 at a forming temperature of 350 °C, a strain rate of 1 s^{-1}, and a logarithmic strain of approximately 0.2. **c** SEM image of the AZ91 alloy showing the precipitation of Mg17Al12. **d** The magnification of precipitation inside the grains and the result of the EDX measurement are shown in **c** [11]

be attributed to the higher aluminum content in AZ91, which encourages the development of dendrite arms [12].

5.3 Mechanical Properties of Magnesium Alloys

Magnesium alloys have a high relative strength [4]. The specific stiffness is approximately the same as that of aluminum and steel but much higher than that of engineering plastics. The specific strength of ultra-light and high-strength Mg-14Li-7Al (LA147, wt.%) is approximately 350 Nm/kg, which is unparalleled among other structural alloys [13]. Under the same harsh conditions, the strength of 1 kg of magnesium alloy is almost equal to 1.8 kg of aluminum or 2.1 kg of steel, and the use of magnesium alloy material for the seat, steering wheel, vehicle cross beam (CCB), and other parts greatly improves the comfort and handling of the car [14].

Magnesium alloys have high damping properties, and the main mechanism is dislocation damping, which is typically 15 times higher than aluminum alloys and 60 times higher than steel. For example, at a stress level of 35 MPa, the Mg damping coefficient of the AZ91D alloy is 25%, and that of the A380 aluminum alloy is only 1%. At a stress level of 100 MPa, the damping coefficients of magnesium alloys AZ91D, AM60, and AS41 are 53, 72, and 70%, respectively, while it is only 4% for A380 aluminum alloy [4].

In addition, magnesium alloys significantly reduce dimensional instability caused by changes in environmental temperature and time compared to other metals. The overall thermal conductivity and heat dissipation properties of magnesium alloy are better than those of aluminum alloy.

Table 5.3 compares the mechanical properties of different classes of magnesium alloys.

The advantages of magnesium as a structural material are summarized in Table 5.4. Some main advantages include specific strength, stiffness, ductility, hot formability, machinability, and damping.

Table 5.3 Comparison of mechanical properties of magnesium alloys [3]

Performance	Mg	AZ91	AM50	AS41	AE42	AISI8Cu3
Density, g/cm^3	1.74	1.81	1.77	1.77	1.79	2.75
Specific heat, kJ/kg K	1.025	1.02	1.02	1.02	1.02	0.88
Thermal conductivity, W/m K	154	51	65	68	84	120
Melting temperature, °C	650	598	620	638	625	593
Solid–liquid conversion temperature, °C	650	420–435	420–435	420–435	590	521
Conversion heat, kJ/kg	368	370	370	370	370	500
Tensile strength, MPa	165–205	240–250	210–230	215–240	230	331
Yield limit, MPa	69–105	160	125	140	145	165
Elongation rate, %	8–11.5	3–7	10–15	6–15	10–11	3
Elastic modulus, GPa	45	45	45	45	45	71
Brinell hardness	36	70	60	60	60	80

Table 5.4 Nominal compositions and typical room-temperature mechanical properties of magnesium alloys

Alloy	Nominal composition, weight percent						Yield strength						
							UTS	Tensile	Compressive	Bearing	Elongation	Shear strength	Hardness
	Al	Mn	Th	Zn	Zr	Other	MPa	MPa	MPa	MPa	%	MPa	HRB
Sand casting and permanent mold casting alloys													
AZ81A-T4	7.6	0.13	–	0.7	–	–	275	83	83	305	15	125	55
AZ91C-T4	8.7	0.13	–	0.7	–	–	275	145	145	360	6	145	66
EQ21A-T6	–	–	–	–	0.7	1.5 Ag 2.1 Dy	235	195	195	–	2	–	65–85
EZ33A-T5	–	–	–	2.7	0.6	3.3 RE	160	110	110	275	2	145	50
HK31A-T6	–	–	3.0	–	0.7	–	220	105	105	275	8	145	55
WE54A-T6	–	–	–	–	0.7	5.2 Y 3.0 RE	250	172	172	–	2	–	75–95
ZC63A-T6	–	0.50	–	6.0	–	2.7 Cu	210	125	–	–	4	–	55–65
ZK61A-T6	–	–	–	6.0	0.7	–	310	195	195	–	10	180	70
High pressure die casting alloys													
AM60	6.0	0.13	–	–	–	–	205	115	115	–	6	–	–
AS21	1.7	0.4	–	–	–	1.1 Si	240	130	130	–	9	–	–
AZ91D	9.0	0.13	–	0.7	–	–	230	150	165	–	3	140	63

(continued)

Table 5.4 (continued)

Alloy	Nominal composition, weight percent						Yield strength						
							UTS	Tensile	Compressive	Bearing	Elongation	Shear strength	Hardness
	Al	Mn	Th	Zn	Zr	Other	MPa	MPa	MPa	MPa	%	MPa	HRB
Extruded bars and shapes													
AZ31B-F	3.0	–	–	1.0	–	–	260	200	97	230	15	130	49
AZ61A-F	6.5	–	–	1.0	–	–	310	230	130	285	16	140	60
ZK60A-T5	–	–	–	5.5	–	–	365	305	250	405	11	180	88
Sheet and plate													
AZ31B-H24	3.0	–	–	1.0	–	–	290	220	180	325	15	160	73
HK3A-H24	–	–	3.0	–	0.6	–	255	200	160	285	9	140	68
HM21A-T8	–	0.6	2.0	–	–	–	235	170	130	270	11	125	–

5.4 Application

AZ91D, AM60B, and AM50 are the most common in automotive Mg alloy applications. Table 5.6 provides a comprehensive overview of the mapping relationships of different families of Mg alloys for automotive components. The producer of the Mg alloy component and applications on the car models are given in Table 5.5.

Some of the specific automotive components produced from magnesium alloys are given in Fig. 5.4.

5.4.1 Body System

Magnesium alloy materials are ideal for thin-walled structural parts and large internal structural parts due to their excellent casting flow properties:

- The advantage of magnesium alloy as a material for the interior lining of doors is that it is light, generally chooses a tensile strength of more than 220 MPa, and has an elongation of more than 10% (Fig. 5.5).
- Replacing hydraulically formed steel, tubular steel, extruded aluminum, and plastic composites with injected magnesium alloys for the front load-bearing components helps reduce additional variable costs (Fig. 5.6).
- The lightweight high-magnesium body structure in the fuel-efficient vehicle program can reduce weight by more than 40% compared to conventional steel structures by replacing the magnesium alloy roof panel (Fig. 5.7).

5.4.2 Chassis System

The use of magnesium alloys in the chassis presents a major challenge for vehicle weight reduction and safety, while strength, ductility, and corrosion resistance must be considered:

- Compared to aluminum discs, magnesium alloy discs have higher strength and toughness. Aesthetics, shock absorption, and heat dissipation are the main advantages of magnesium alloy wheels (Fig. 5.8).
- The application of magnesium alloy in steering wheels is widely accepted worldwide due to its energy-absorbing and vibration-dampening capabilities. A magnesium alloy steering wheel offers performance benefits that traditional steering wheels, such as mild steel and engineering plastics, cannot match (Fig. 5.9).
- A magnesium alloy subframe is the most effective way to achieve a lightweight car chassis structure, which also has the advantage of isolating road vibrations and improving stability. In addition, the small number of assembled parts makes installation easier and lowers assembly costs (Fig. 5.10).

Table 5.5 Producers of Mg alloy components and applications on car models [15]

Component	Producers and car models
Engine block	BMW: a lighter, more powerful, and more durable six-cylinder inline combustion engine. The world's first engine block made of Noranda's patented alloy AJ62 (Mg-Al-Sr)
Steering wheel frame	Ford (Ford Thunderbird, Cougar, Taurus, Sable), Chrysler (Chrysler Plymouth), Toyota, BMW (MINI), Lexus (Lexus LS430)
Seat frame	GM (impact), Mercedes-Benz (Mercedes Roadster 300/400/500 SL), Lexus (Lexus LS430)
Instrument panel	GM, Chrysler (jeep), Ford, Audi (A8), Toyota (Toyota Century)
Wheel rims	Toyota (Toyota 2000GT, Toyota Supra), Alfa Romeo (GTV), Porsche AG (911 Serie)
Cylinder head	Dodge (Dodge Raw), Honda Motor (City Turbo), Alfa Romeo (GTV), AutoZAZ-Daewoo (Tavria, Slavuta, Daewoo-Sens), Honda, BMW, Ford, Isuzu, Volvo Motors (LCP), Chrysler
Clutch case	AutoZAZ-Daewoo (Tavria, Slavuta, Daewoo-Sens), Volvo Motors (LCP), Alfa Romeo (GTV)
Transmission case	AutoZAZ-Daewoo (Tavria, Slavuta, Daewoo-Sens), Volvo Motors (LCP), Porsche AG (911 Serie), Volkswagen (Volkswagen Passat), Audi (A4,A6), Mercedes-Benz
Lower crankcase	Chrysler (jeep), Alfa Romeo (GTV), GM (Oldsmobile), McLaren Motors (F1-V12)
Cylinder block (without liners and main bearing heads)	GM (Pontiac Gran AM, Corvette)
Intake manifold	GM (V8 North Star motor), Chrysler
Air intake system	BMW (V8 motor)
Steering link bracing	GM (LH Midsize)
Oil pump body	McLaren Motors (F1-V12)
Camshaft drive chain case	Porsche AG (911 Serie)
Gear controls housing	AutoZAZ-Daewoo (Tavria, Slavuta, Daewoo-Sens)
Brackets for air comfort system compressor, steering booster pump, and generator	Chrysler, Volkswagen (Volkswagen Lupo)

Table 5.6 Application areas of Mg alloy series in car body [15]

Mg alloy series	Performance advantages	Specific classification	Applications	Die casting process
AZ (Mg-Al-Zn-Mn)	Application to high strength, low ductility components in thermal environments below 125 °C	AZ91D	Central control bracket, transmission housing, oil pan, oil filter housing, cylinder header cover, steering column bracket, inner panel, clutch housing, swivel plate, starter housing, headlight bracket	Gravity or low pressure Sand or metal mold High-pressure die-casting Squeeze casting Thixocasting
		AZ61	Luggage rack skeleton, column beam	
		AZ31	Inner panel, compressor housing	
AM (Mg-Al-Mn)	Used in thermal environments up to 125 °C for lower strength but more ductile components	AM50	Seat frame, wheel/hub, inside plate, steering wheel	Gravity or low pressure Sand or metal mold High-pressure die-casting Squeeze casting Thixocasting
		AM60B	Front-end frame, cross-car beam, seat frame, steering wheel, wheel/hub, radiator bracket, inside door panel	
AS (Mg-Al-Si-Mn)	Replaces AZ alloys at 125 °C in thermal environments due to increased creep strength	AS41B	Transmission housing, crankcase housing	High-pressure die casting
AE (Mg-Al-E-Mn)	Higher crawling strength in 150 °C drivetrain operating environment, but expensive due to E content	AE44	Transmission case, oil pan, engine bracket	High-pressure die casting
AXE and AJ (Mg-Al-Sr/Ca-Mn)	Applications for high creep strength powertrain components, but with potential low cost compared to AE	AJ52	Engine components	High-pressure die casting
		AJ62	Engine block	
		AXJ530	Engine components	

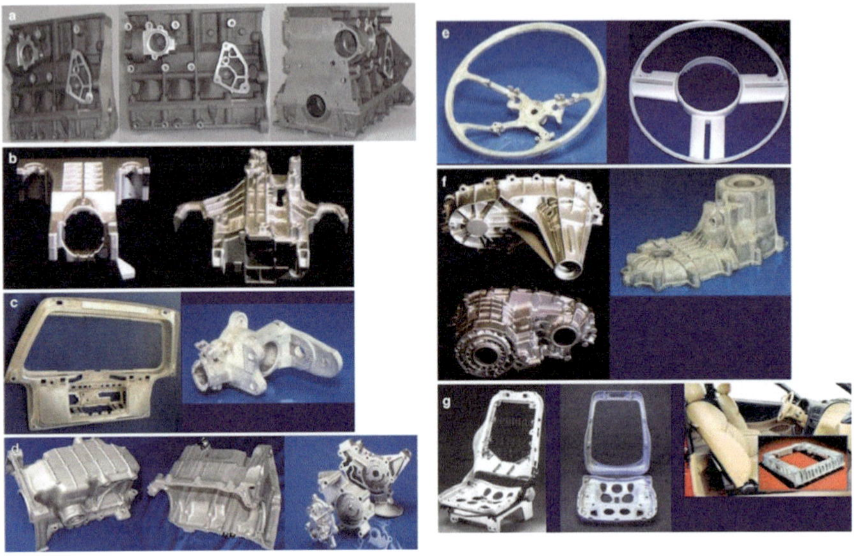

Fig. 5.4 Examples of automotive components made of Mg alloys: **a** engine block, **b** steering column module, **c** door frame/key lock housing, **d** oil pan, **e** steering wheel, **f** transfer case/transmission housing, **g** seat frame, **h** wheel [15]

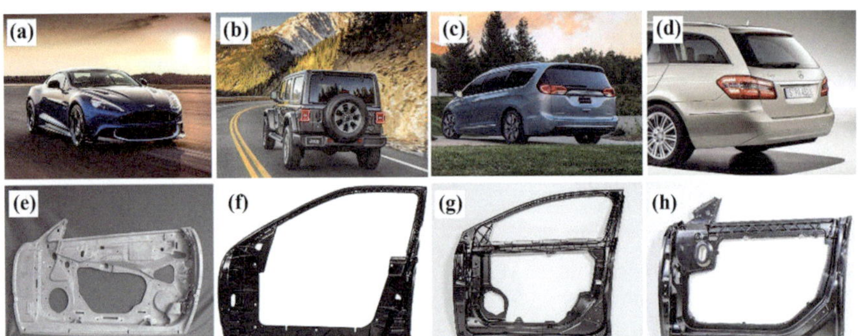

Fig. 5.5 Development and application of Mg alloys in car doors: **a** Aston Martin Vanquish S with cast Mg side door inner; **b** all-new 2018 Jeep Wrangler produced with a die-cast Mg rear swing gate; **c** Chrysler Pacifica showing the liftgate assembly highlighted by a Mg die-cast inner; **d** the rear end of the Mercedes-Benz E-Class T-model featuring the hybrid Mg-Al hatchback; **e** inner door frame of the Daimler-Chrysler SL Roadster; **f** Ford's concept die-cast Mg door inner with an open architecture; **g** integrated Mg die-cast door inner designed as part of a DOE-sponsored project led by GMC; **h** ultra-thin and ultra-light Mg alloy door inner [4]

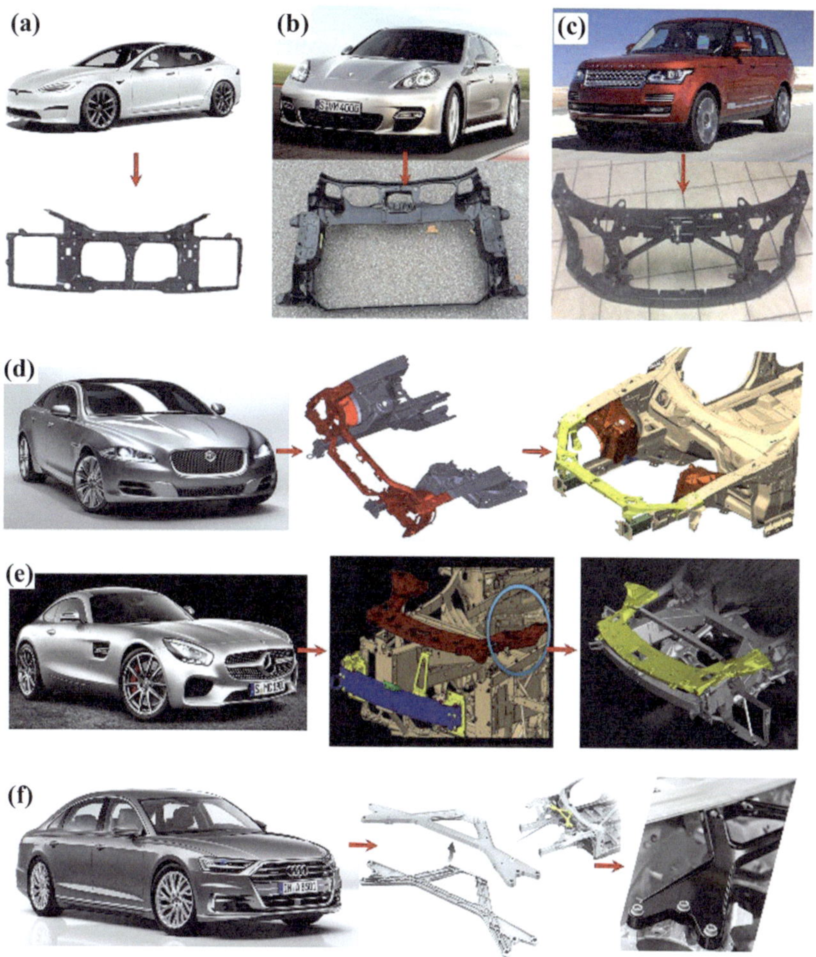

Fig. 5.6 Development and application of Mg alloys in front-end carrier and front upper component: **a** Tesla model S one-piece die-cast Mg front-end carrier; **b** Porsche Panamera G2 front-end carrier; **c** Range Rover Mg alloy front-end carrier; **d** Jaguar XJ Mg alloy front upper component; **e** Mercedes-benz AMG GT Mg alloy front upper component; **f** Audi A8 Mg cabin bracket [4]

5.4.3 Powertrain System

Due to the superior casting properties of magnesium alloys, engine blocks, cylinder heads, transmissions, and oil pans begin to change from the original aluminum alloy material:

- VW and Audi are equipped with Mg engines and transmissions whose materials work at temperatures below 120 °C, which is Mg's most important weight reduction capability (Fig. 5.11).

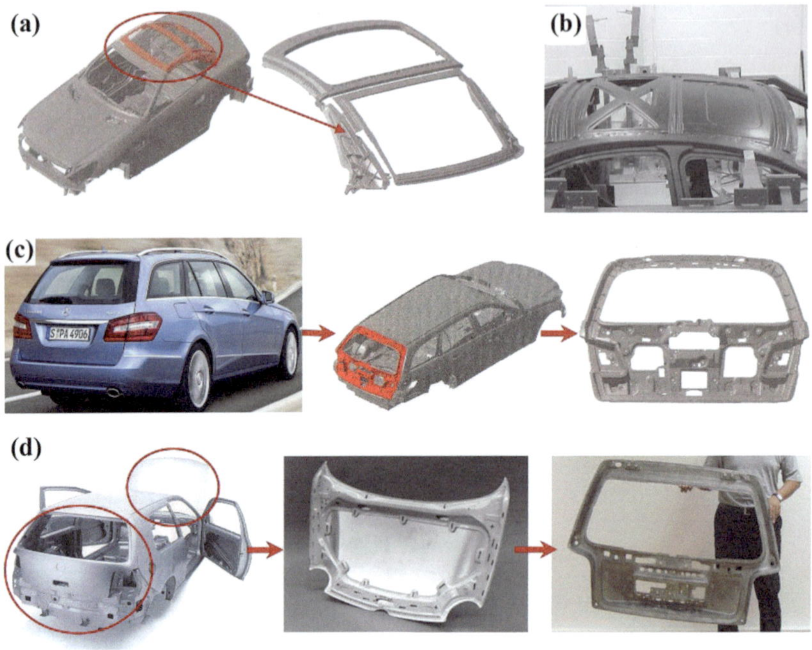

Fig. 5.7 Development and application of Mg alloy on roof, hood, and trunk lid: **a** Mg alloy folding roof for Mercedes-Benz SL/SLK series cars; **b** Mg alloy inner plate from Daimler-Chrysler; **c** Mercedes-Benz E-Class touring cassenger car trunk lid; **d** VW Lupo Mg trunk lid and hood [4]

- Currently, most oil pans are made of aluminum alloy. The trend of reducing vibration and noise has led to the need to find a lighter material, and in this regard, Mg alloys are preferred due to their advantages (Fig. 5.12).

5.4.4 Internal System

Magnesium alloy car interior parts must have good strength, ductility, energy absorption and casting properties regarding vehicle safety. Some typical interior systems include seat frames (Fig. 5.13), CCBs (Fig. 5.14), center console brackets (Fig. 5.15), steering covers, etc.

5.5 Summary

Magnesium alloys, composed of elements like aluminum, zinc, and rare earth metals, offer advantages such as enhanced strength, corrosion resistance, and excellent castability. The historical use of magnesium in automobiles dates to the early 20th century

Fig. 5.8 The development and application of Mg alloy in the wheel: **a** Mg alloy wheel for Chevrolet Corvette; **b** Cadillac CT4-V forged spun Mg wheel; **c** AMG Project One 9-spoke Mg forged wheels with bionic design; **d** Hollow billet extruded Mg alloy wheel: (14 × 6) J; © Hollow billet extruded Mg alloy wheel: (13 × 10) J; **f** Hollow billet extruded Mg alloy wheel: (13 × 8) J; **g** Bugatti Chiron Super Sport 300 + Mg wheel; **h** Porsche 911 GT3 RS Mg forged wheel; **i** Bandit9 electric racing Mg wheel; **j** Mg alloy car wheel forward and reverse extrusion forming technology process, from Dingxin Magnesium Technology Co [4]

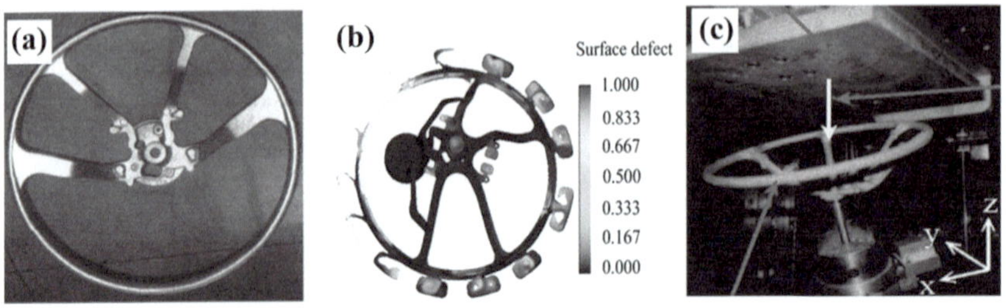

Fig. 5.9 The development and application of Mg alloy in the steering wheel: **a** Mg alloy steering wheel frame produced by Chongqing Mg Industry; **b** Surface defect probability simulation result; **c** Mg alloy steering wheel crash test [4]

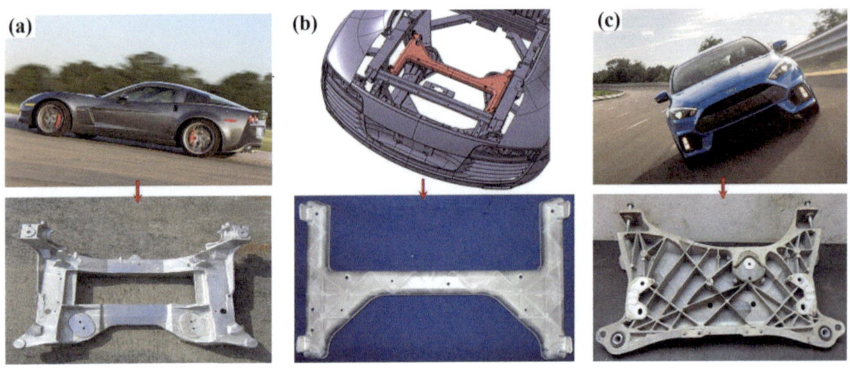

Fig. 5.10 The development and application of Mg alloy in subframe: **a** Corvette Z06 Mg subframe; **b** Audi R8 Mg subframe; **c** 2016 Ford Focus Mg subframe [4]

and its expansion into aerospace and military applications. Despite challenges like high-temperature creep and corrosion, continuous alloy development has led to improved performance and increased adoption in automotive components.

Magnesium alloys such as AZ91, AM50, and AM60 are widely used due to their balance of strength, ductility, and corrosion resistance. The mechanical characteristics of magnesium alloys, including high specific strength, damping properties, and thermal conductivity, make them ideal for automotive applications. Comparisons with other metals demonstrate magnesium's advantages in reducing vehicle weight and improving efficiency.

Magnesium alloys are extensively applied in different vehicle systems. In the body system, magnesium components like doors, hoods, and trunk lids contribute to weight reduction while maintaining structural integrity. In chassis systems, magnesium wheels, and subframes enhance vehicle stability and performance. The powertrain system benefits from magnesium alloy engine blocks, cylinder heads, and oil pans, reducing vehicle mass while maintaining strength. Additionally, interior systems such as seat frames, center console brackets, and steering wheels utilize magnesium alloys for their durability and energy absorption properties.

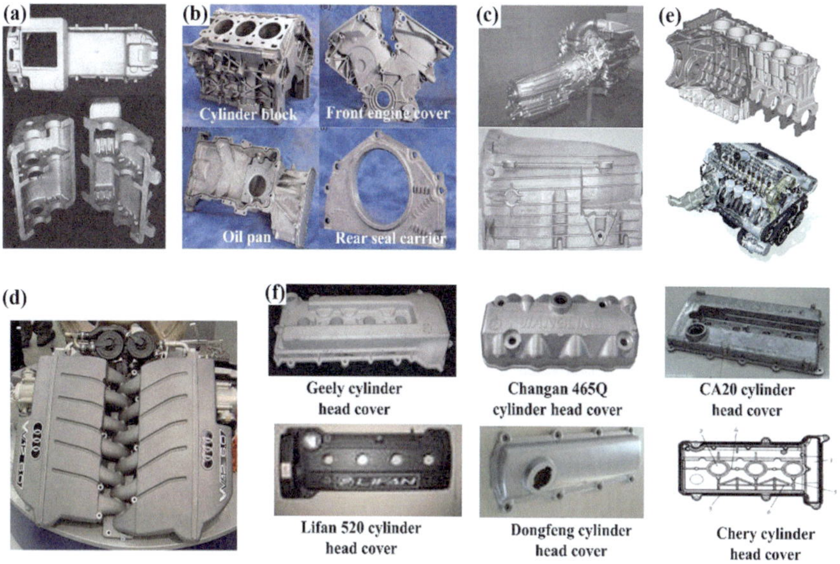

Fig. 5.11 Mg alloy in powertrain applications: **a** Mg applications in the 1930s: crankcase and transmission housing; **b** Mg powertrain components from the USAMP Mg powertrain cast components project; **c** The Mercedes 7-speed automatic transmission case; **d** Audi A8 (12-cylinder) intake manifold cover; **e** BMW 6-cylinder R6 engine; **f** engine cylinder head cover typical cases [4]

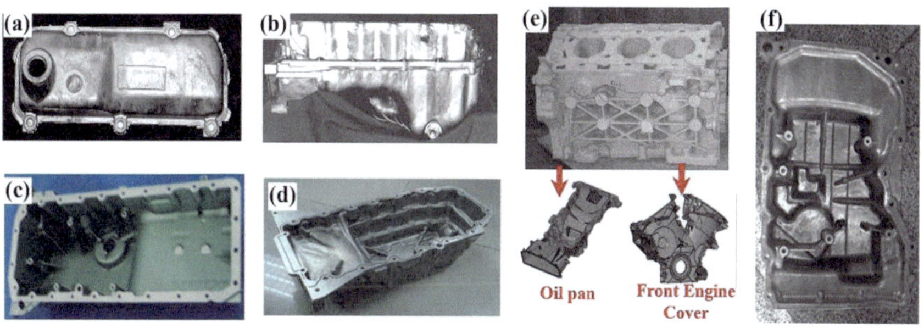

Fig. 5.12 Development and application of Mg alloy in oil pan: **a** valve-cover cast in AJ52x at Spartan light metal products with very good results; **b** oil pan cast in AJ52x at Spartan light metal products with very good results; **c** Honda Insight Mg alloy oil pan; **d** CA20 Mg alloy oil pan; **e** USAMP Mg engine oil pan and engine cover; **f** high-temperature creep resistant die-cast Mg alloy oil pan for FAW Beston car [4]

While magnesium remains the third most-used metal after steel and aluminum, challenges such as load-bearing limitations and corrosion resistance continue to be addressed through advanced manufacturing techniques and alloy improvements. The

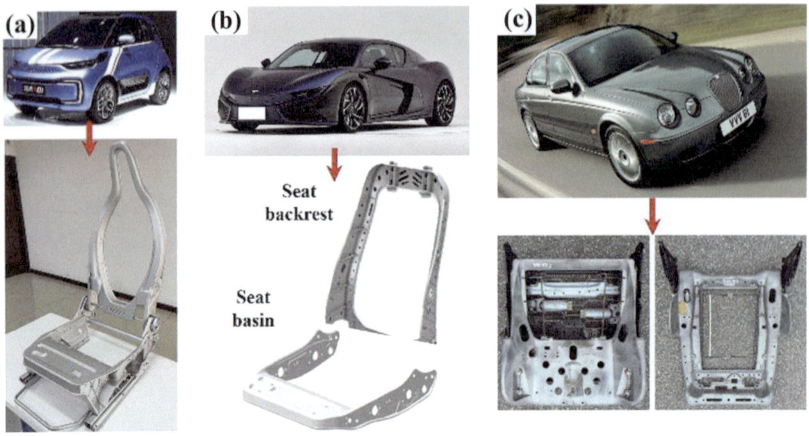

Fig. 5.13 Application of Mg alloy on seat frame: **a** Changan E01 Mg alloy seat frame; **b** future K50 Mg alloy seat frame; **c** Jaguar S Type's Mg alloy seat frame [4]

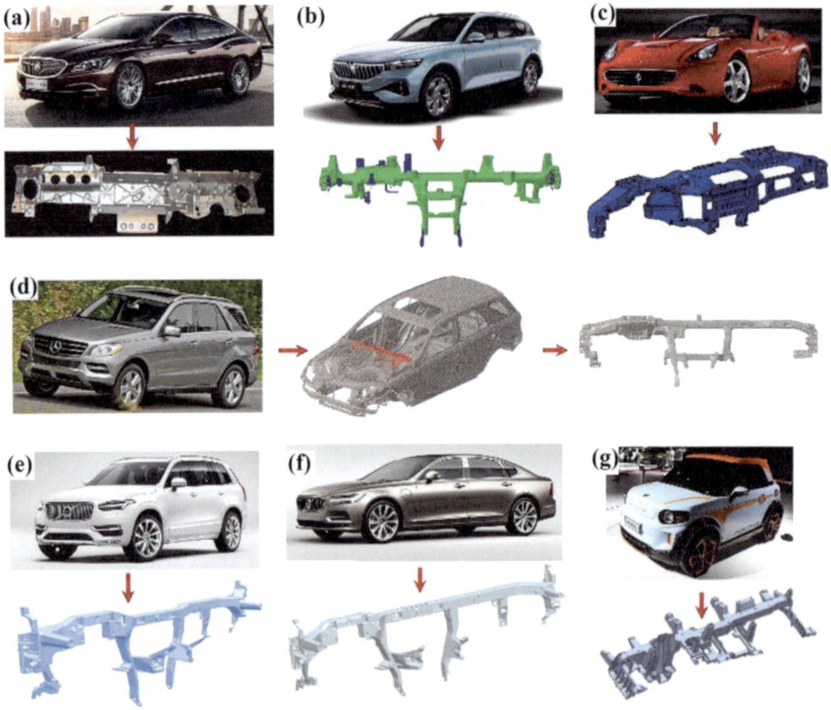

Fig. 5.14 Several applications of Mg alloy CCB: **a** Buick grand touring die-cast Mg alloy CCB; **b** Mg alloy CCB on the Voyah FREE model; **c** Ferrari California die-cast Mg alloy CCB; **d** Mercedes-Benz M series die-cast Mg alloy CCB; **e** Volvo new XC90 K426 die-cast Mg alloy CCB; **f** Volvo new S90 L541 die-cast Mg alloy CCB; **g** BAW ARCFOX-1 die-cast Mg alloy CCB [4]

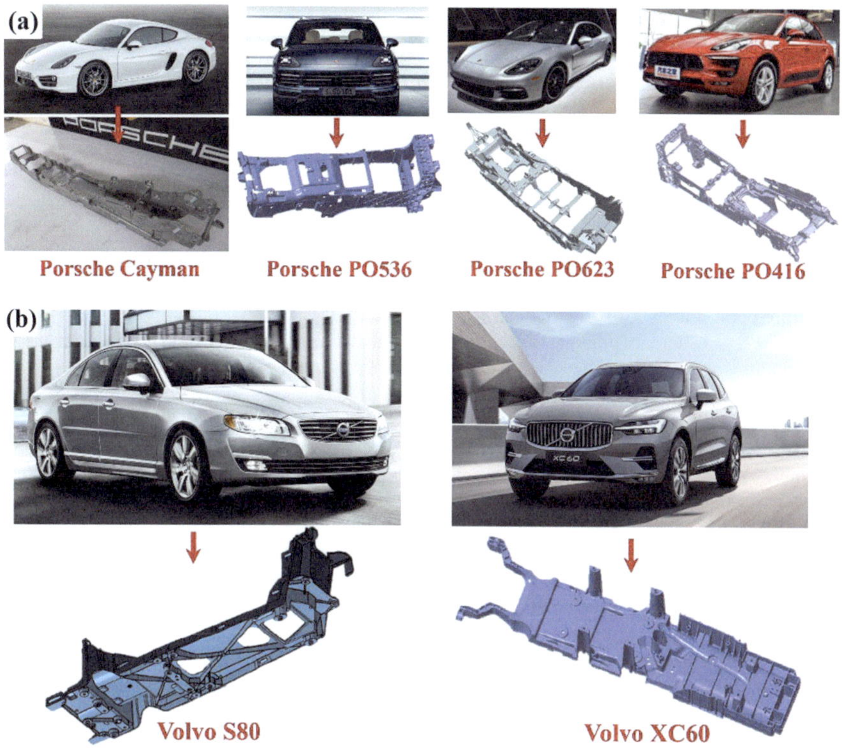

Fig. 5.15 Several Mg alloy center bracket applications: **a** Porsche series Mg alloy center bracket; **b** Volvo series Mg alloy center bracket [4]

widespread adoption of magnesium alloys is driven by the need for lightweight, fuel-efficient vehicles, positioning magnesium as a key material in the future of automotive engineering.

References

1. Unnikrishnan MA, Raja Dhas JE, Savio Lewise KA, Varghese JC, Ganesh M (2023) Challenges on friction stir welding of magnesium alloys in automotives. Materials Today Proc. https://doi.org/10.1016/j.matpr.2023.03.789
2. Satya Prasad SV, Prasad SB, Verma K, Mishra RK, Kumar V, Singh S (2022) The role and significance of magnesium in modern day research-a review. J Magnes Alloys 10(1):1–61. https://doi.org/10.1016/j.jma.2021.05.012
3. Taub AI, Luo AA (2015) Advanced lightweight materials and manufacturing processes for automotive applications. MRS Bull 40:1045–1054. https://doi.org/10.1557/mrs.2015.268
4. Liu B, Yang J, Zhang X, Yang Q, Zhang J, Li X (2023) Development and application of magnesium alloy parts for automotive OEMs: a review. J Magnes Alloys 11(1):15–47. https://doi.org/10.1016/j.jma.2022.12.015

5. Tan J, Ramakrishna S (2021) Applications of magnesium and its alloys: a review. Appl Sci 11:6861. https://doi.org/10.3390/app11156861
6. Sankaranarayanan S, Gupta M (2021) Emergence of god's favorite metallic element: magnesium based materials for engineering and biomedical applications. Mater Today: Proc 39(1):311–316. https://doi.org/10.1016/j.matpr.2020.07.220
7. Loukil N (2022) Alloying elements of magnesium alloys: a literature review. In: Magnesium alloys structure and properties. IntechOpen, UK https://doi.org/10.5772/intechopen.96232
8. Murugesan R et al (2023) Influence of alloying materials Al, Cu, and Ca on microstructures, mechanical properties, and corrosion resistance of mg alloys for industrial applications: a review. ACS Omega 8(41):37641–37653. https://doi.org/10.1021/acsomega.3c03417
9. Wu T, Zhang K (2023) Corrosion and protection of magnesium alloys: recent advances and future perspectives. Coatings 13:1533. https://doi.org/10.3390/coatings13091533
10. ASTM B 951 standard practice for codification of unalloyed magnesium and magnesium-alloys, cast and wrought. American Society for Testing and Materials (2018)
11. Moses M, Ullmann M, Prahl U (2023) Influence of aluminum content on the microstructure, mechanical properties, and hot deformation behavior of Mg-Al-Zn alloys. Metals 13:1599. https://doi.org/10.3390/met13091599
12. Dahle AK, Lee YC, Nave MD, Schaffer PL, StJohn DH (2001) Development of the as-cast microstructure on magnesium-aluminium alloys. J Light Met 1:61–72. https://doi.org/10.1016/S1471-5317(00)00007-9
13. Xin T et al (2021) Ultrahigh specific strength in a magnesium alloy strengthened by spinodal decomposition. Sci Adv 7:eabf3039. https://doi.org/10.1126/sciadv.abf3039
14. Li J, Chen L, Jiang S, Gan H, Hao W (2023) Numerical simulation and process optimization of magnesium alloy vehicle dashboard cross car beam (CCB) based on MAGMA. Processes 11:16. https://doi.org/10.3390/pr11010016
15. Kulekci MK (2008) Magnesium and its alloys applications in automotive industry. Int J Adv Manuf Technol 39:851–865. https://doi.org/10.1007/s00170-007-1279-2

Chapter 6
Copper and Its Alloys

Abstract The chapter explores the critical role of copper and its alloys in the automotive industry, highlighting their widespread use due to outstanding electrical, thermal, mechanical, and corrosion-resistant properties. As vehicle technologies evolve toward electrification and intelligence, demand for high-performance copper alloys—such as brass, bronze, copper-nickel, and specialized variants—continues to grow. These materials are essential in systems such as power electronics, braking, transmissions, and thermal management. The chapter details alloy classifications per standard compositions and mechanical properties. Key materials are discussed for their superior strength and conductivity. The chapter concludes by illustrating alloy applications across various products.

6.1 Introduction

The automotive industry has become one of the important applications of copper alloy materials. With the development of automobiles towards electrification and intelligence, higher demands are placed on the complex performance characteristics of copper alloy materials [1].

High-performance copper alloys are widely used in heat dissipation systems, electronic power systems, brake systems, and hydraulic systems of automobiles [2]. They are key basic raw materials for important automotive components.

According to the statistics, 23 kg of copper is consumed by internal combustion engines of traditional vehicles, 40 kg by hybrid electric cars, 60 kg by plug-in hybrid electric cars, and 83 kg by electric cars [3]. With the continuous development of new energy vehicles, the demand for copper in the automotive industry will increase.

The charging industry, which supports electric vehicles, also has a high demand for copper. The main power equipment of the charging station is power cables and transformers, as well as circuit breakers, fuses, various switches, connectors, etc. The main copper components of a charging station include charging cables, charging modules, connectors, and various switches [3].

K. Berladir et al., *Materials for the Automotive Industry*, Springer Tracts in Mechanical Engineering, https://doi.org/10.1007/978-3-032-05866-9_6

6.2 Classification of Copper Alloys

Copper alloys play a key role in the automotive industry. They are a diverse range of metal compounds consisting primarily of copper and other elements such as zinc, aluminum, nickel, and tin. These alloys are characterized by exceptional mechanical properties, including high strength, corrosion resistance, and thermal conductivity, making them irreplaceable in the automotive industry. Copper alloys for the automotive market include brass, bronze, and copper-nickel alloys that perform critical functions in manufacturing components such as heat sinks, connectors, terminals, and electrical systems [4].

Copper and copper alloys are labeled according to the Copper Development Association (CDA) [5]. In the designation system, the numbers C10000 to C79999 indicate wrought alloys. Cast alloys are numbered C80000 to C99999. Within these two categories, the compositions are grouped into the following groups: copper and copper alloys (Table 6.1).

The composition limits in this list of standard designations for copper and copper alloys are given by the following numerical convention where possible (Table 6.2).

Due to the cost of materials and the immobilization of vehicle materials, mainly brass, bronze, and some C70250 alloys are currently used for automotive components.

6.3 Properties of Copper Alloys

Copper alloys have become an integral part of the automotive industry due to their exceptional properties and versatility in metalworking. These alloys offer many advantages, including high thermal and electrical conductivity, excellent corrosion resistance, and superior strength. From electrical components to engine parts, copper alloys have numerous applications in modern vehicles, increasing performance, efficiency, and longevity [2, 6–8].

Electrical properties. Due to their exceptional conductivity, copper alloys are the number one choice for electrical components. Copper is a highly conductive metal by itself, and when alloyed with other elements such as tin or nickel, it further improves its electrical properties. For example, copper-nickel alloys, commonly known as cupronickel, are widely used in automotive wiring harnesses, connectors, and clamps. These alloys offer excellent corrosion resistance and reliable electrical connections critical to various vehicle systems.

Thermal properties. Regarding thermal management, copper alloys play an important role in the automotive industry. Engine cooling systems rely heavily on copper-based heat exchangers and radiators. The high thermal conductivity of copper ensures efficient heat transfer, optimal engine performance, and the prevention of overheating. In addition, copper alloys are also used to manufacture heat sinks for electronic components, further improving thermal management in modern vehicles.

Table 6.1 The groups of the copper alloys [5]

Copper alloy	Explanation
Coppers	Metals with a specified minimum copper content of 99.3% or higher
High copper alloys	For wrought products, these are alloys with a specified copper content of less than 99.3% but more than 96%, which do not fall into another group of copper alloys. Cast copper alloys have a specified copper content above 94%, to which silver may be added for special properties
Brasses	These alloys use zinc as the main element, with or without adding other alloying elements such as iron, aluminum, nickel, and silicon Wrought alloys include three main groups of brasses Copper-zinc alloys Copper-zinc-lead alloys (leaded brasses) Copper-zinc-tin alloys (tin brasses) The cast alloys comprise four main families of brasses Copper-tin-zinc alloys (red, semi-red, and yellow brasses) "Manganese bronze" alloys (high-strength yellow brasses) Leaded "manganese bronze" alloys (leaded high-strength yellow brasses) Copper-zinc-silicon alloys (silicon brasses and bronzes) Cast copper-bismuth and copper-bismuth-selenium alloys
Bronzes	Bronzes are copper alloys, the major alloying element of which is not zinc or nickel. Originally it described alloys with tin as the only or principal alloying element For wrought alloys, there are four main families of bronzes Copper-tin-phosphorus alloys (phosphor bronzes) Copper-tin-lead-phosphorus alloys (leaded phosphor bronzes) Copper-aluminum alloys (aluminum bronzes) Copper-silicon alloys (silicon bronzes) The cast alloys have four main families of bronzes Copper-tin alloys (tin bronzes) Copper-tin-lead alloys (leaded and high leaded tin bronzes) Copper-tin-nickel alloys (nickel-tin bronzes) Copper-aluminum alloys (aluminum bronzes)
Copper-nickels	These are alloys with nickel as the principal alloying element, with or without other designated elements
Copper-nickel-zinc alloys	Known commonly as "nickel silvers," these alloys contain zinc and nickel as the principal and secondary alloying elements, with or without other designated elements
Leaded coppers	These comprise a series of cast alloys of copper with 20% or more lead, sometimes with a small amount of silver, but without tin or zinc
Special alloys	Alloys whose chemical compositions do not fall into the above categories are combined into "special alloys"

Corrosion resistance. Corrosion resistance is another critical factor in the automotive industry, especially in harsh environments. Copper alloys such as brass and bronze exhibit excellent corrosion resistance, making them ideal for various applications. For example, brass is commonly used in fittings and connectors for fuel lines and brake systems where exposure to moisture and corrosives is constantly

Table 6.2 Standard chemical composition limits of copper alloys [6]

Constituent	Convention for expressing composition limits
Copper Nos. C10000–C15999 (wrought) and C80000–C81199 (cast)	
Copper	XX.XX
Alloying elements and impurities	
1/10 and over	.XX
1/100 and less than 1/10	0.0XX
1/1000 and less than 1/100	0.00X
Less than 1/1000	0.000X
Copper Alloy Nos. C16000–C79999 (wrought) and C81300–C99999 (cast)	
Copper	XX.X
Alloying elements and impurities	
6/10 and over	.X
1/10 and less than 6/10	.XX
1/100 and less than 1/10	0.0X
1/1000 and less than 1/100	0.00X
Less than 1/1000	000X

challenging. Its corrosion resistance ensures the durability and reliability of these components.

Mechanical properties. Strength and durability are important factors in the automotive industry. Copper alloys have excellent mechanical properties, making them suitable for various structural components. For example, beryllium copper alloys are known for their high strength, wear, and fatigue resistance. These alloys are often used in engine valve seats, valve guides, and bushings, where durability and reliability are paramount. By using copper alloys in these critical areas, car manufacturers can ensure the durability and performance of their vehicles.

When comparing the various automotive copper alloy options, it is important to consider factors such as cost, availability, and specific application requirements. For example, although copper-nickel alloys have excellent corrosion resistance, they can be more expensive than other copper alloys. In such cases, brass or bronze alloys may be a viable alternative that provides satisfactory corrosion resistance at a lower cost. In addition, the availability of specific alloy compositions and their compatibility with manufacturing processes should be considered.

6.4 Types of Copper Alloys

6.4.1 Complex Brass

It is used in structural friction components such as automotive transmissions where synchronizers are critical parts connecting input and output shaft gears. It must meet the following performance criteria [7]:

- Outstanding mechanical qualities.
- Steady friction coefficient.
- Good wear resistance.
- No wear on the dual cone surface.
- Good oil compatibility.
- Good heat resistance/thermal conductivity.
- Simple processing.

6.4.2 Dispersion Strengthened Copper

Resistance welding joins workpieces, applies pressure through electrodes, and uses resistance heat generated by current passing through the contact area of the joint and the surrounding area to make the weld. Each car body has approximately 4000–6000 connections made by contact soldering.

The requirements for resistance welding electrode materials include:

- High-temperature hardness and strength.
- High oxidation resistance.
- Low tendency to form a copper alloy with weldment materials.
- Suitable electrical and thermal conductivity.

Dispersion-strengthened copper alloy meets these requirements. It is a high-strength, highly conductive, and highly heat-resistant copper alloy with the best comprehensive physical and mechanical properties [9, 10].

Compared with the traditional internal oxidation process, this new production process with a short process is an original new «reaction synthesis» technology. It uses a new type of oxygen source to process "reaction synthesis" to ensure the completeness of the reaction and obtain pure γ-Al_2O_3 nanoparticles, eliminating the «internal oxidation method» in the impurity phase of the solid oxygen source.

6.4.3 Copper-Chromium Series

Typical age-strengthened alloys use chromium as the main strengthening phase, which has:

- High electrical conductivity.
- Strength.
- Stress relaxation resistance.

Copper-chromium alloys can be aged and hardened with judicious techniques to improve bending and strength. High-performance copper alloys are widely used in heat dissipation systems, electronic power systems, brake systems, and hydraulic systems of automobiles.

In typical age-strengthened alloys, the BCC chromium phase is the main strengthening phase of the alloy, which has high electrical conductivity, good strength, and good resistance to stress relaxation. Countries worldwide have successfully conducted theoretical and industrial research on Cu-Cr alloys [11–13].

Technology of bending and control of mechanical properties of copper and chromium alloys. The strengthening mechanism of copper-chromium alloys mainly depends on aging and mechanical strengthening. Judicious coordination of strengthening methods will help control bending and strength.

The technology of substituting zirconium elements, depending on the field of application, comprehensively investigates the influence of the synergistic effect of alloying elements on the complex properties of copper and chromium alloys and selects reasonable substitution elements.

6.4.4 C70250 High-Performance Alloy

Alloy C70250 (CuNi3Si1) is a high-performance alloy that combines strength, conductivity, ductility, and stress relief into a unique set of properties [14]. C70250 has the properties of some non-beryllium beryllium coppers (Table 6.3).

The alloy is a thermally aged material. It achieves its properties through a combination of cold and heat processing, all carried out in the factory. The alloy's high strength, conductivity, ductility, and stress-relief properties make C70250 an excellent electronic alloy, especially in high-temperature environments (Tables 6.4 and 6.5).

Table 6.3 Chemical composition of alloy C70250, wt. %

Ni*	Si	Mg	Zn	Pb	Fe	Mn	Cu**
2.2–4.2	0.25–1.2	0.05–0.30	Max 1.0	Max 0.05	Max 0.20	Max 0.10	Min 99.5

* Including Co; ** including named elements

Table 6.4 Physical properties of alloy C70250 (in annealed temper at 20 °C)

Parameter	Value
Density, g/cm^3	8.8
Thermal expansion coefficient, 10^{-6}/K	17.6
Specific heat capacity, J/(g K)	0.399
Thermal conductivity, W/(m·K)	190
Electrical conductivity, MS/m	23
Electrical conductivity, %	40
Thermal coefficient of electrical resistance, 10^{-3}/K	3
Modulus of elasticity, GPa	130

Table 6.5 Mechanical properties of alloy C70250 (EN 1652)

Temper	Tensile strength R_m, MPa	Yield strength minimum $R_{p0.2}$, MPa	Elongation minimum A_{50mm}, %	Hardness, HV
R620	620–760	500	10	180–240
R655	655–825	585	7	190–250
R690	690–860	655	5	220–260

6.4.5 Cu-Mg Copper Alloys

MSP Series of high-performance are solid solution-strengthened Cu-Mg alloys [6]. The lineup includes:

- MSP1 alloy has been well received by customers and has been used for many years in electrical terminals for automotive applications, busbars, and relays;
- MSP5 alloy has higher strength, flexibility, and low specific weight, making it ideal for use in compact terminals;
- MSP8 alloy has excellent electrical conductivity and excellent resistance to stress relaxation, making it ideal for use in high voltage and high current applications such as busbars and high voltage terminals.

MSP1 is widely used as a copper alloy for terminals, and its stable quality, performance, and power supply make it the preferred choice for many users. MSP1 has also been adopted for use in high-voltage terminals of electric vehicles and hybrid electric vehicles.

MSP5 is the strongest copper alloy in the MSP series. The combination of proper electrical conductivity, excellent resistance to stress relaxation, and flexibility makes it suitable for use as a termination material for a wide range of electronic and electrical equipment where compact size is required. It is also suitable as an alternative to Corson alloys and phosphor bronze.

MSP8 is the most suitable copper alloy for high voltage and high current applications in the MSP series. It provides high electrical conductivity (approx. 80% IACS) and excellent resistance to stress relief and is ideal for use as a busbar and high voltage termination material for new-generation vehicles. It also has excellent formability and high strength, which makes it possible to create precise geometries.

The chemical composition and physical properties of MSP series alloys are shown in Tables 6.6 and 6.7.

The mechanical properties of MSP1, MSP5, and MSP8 alloys are shown in Tables 6.8, 6.9, and 6.10, respectively.

The MSP series alloys have sufficient softening resistance for lead and spring material use. They are particularly suitable as an alternative to phosphor bronze. MSP1 has excellent resistance to stress relaxation as a copper alloy and is extremely reliable as a termination material. MSP5 has stress relaxation resistance equal to

Table 6.6 Chemical composition of MSP series alloys

Alloy	Weight, %		
	Mg	P	Cu
MSP1	0.70	0.005	Remainder, including inevitable impurities and trace additive elements
MSP5	1.60	–	
MSP8	0.25	0.002	

Table 6.7 Physical properties of MSP series alloys

Parameter	MSP1	MSP5	MSP8
Specific gravity (297 K)	8.8	8.5	8.9
Coefficient of thermal expansion ($\times 10^{-6}$/K: 293–573 K)	17.3	18.2	17.8
Thermal conductivity (W/(m·K): 293 K)	264	174	340
Volume resistivity (μΩm: 293 K)	0.027	0.042	0.021
Electrical conductivity (%IACS: 293 K)	63	43	82
Modulus of elasticity (kN/mm^2: 293 K)	125	115	131

Table 6.8 Mechanical properties of MSP1 alloy

Parameter	Temper					Typical value			
	1/4H	1/2H	H	EH	SH (min)	1/4H	1/2H	H	EH
Tensile strength (MPa)	365–450	420–510	480–570	540–630	590	399	459	530	585
0.2% yield strength (MPa)	300–410	370–480	440–550	490–620	540	328	432	494	560
Elongation (%)	15 min	10 min	7 min	–	5	25	15	11	8
Vickers hardness (HV)	90–140	120–170	150–190	170–210	180	126	144	162	178

Table 6.9 Mechanical properties of MSP5 alloy

Parameter	Temper				Typical value (LD/TD)			
	1/2H	H	EH	SH	1/2H 0.64 mm thickness	H 0.15 mm thickness	EH 0.15 mm thickness	SH 0.15 mm thickness
Tensile strength (MPa)	485–585	530–630	575–675	620–720	543/565	593/645	625/695	677/780
0.2% Yield strength (MPa)	–	–	–	–	507/512	543/592	585/647	636/735
Elongation (%)	5 min	4 min	3 min	3 min	11/16	9/14	7/12	7/10
Vickers hardness (HV)	145–205	160–220	175–235	190–250	178	193	203	219

Table 6.10 Mechanical properties of MSP8 alloy

Parameter	Temper				Typical value (LD/TD)			
	1/4H	1/2H	H	EH	1/4H 3.0 mm thickness	1/2 0.4 mm thickness	H 0.8 mm thickness	EH 0.64 mm thickness
Tensile strength (MPa)	270–370	320–420	360–460	420–520	325	367	415	459
0.2% Yield strength (MPa)	–	–	–	–	282	331	399	434
Elongation (%)	10 min	6 min	3 min	2 min	17	16	9	9
Vickers hardness (HV)	70–130	84–145	100–160	110–170	102	113	130	141

(or better than) our Corson alloys, making it suitable for use as an end material in automotive and other industrial applications. MSP8 has high electrical conductivity and excellent resistance to stress relaxation, making it suitable for use in high-voltage lines and busbars. MSP1 has excellent fatigue properties as a spring material, and MSP8 is suitable for use as a material for relays (moving parts) and springs because they can be used for many cycles before failure.

Cu-Mg copper alloys (MSP series) are used to produce terminals (automotive industry), relays (moving parts), spring materials for contacts, bars, switches, battery terminals, fuse terminals, compact switches, electrically conductive parts in lithium-ion batteries, junction boxes and relay boxes [2].

6.5 Application

Copper and its alloys play a crucial role in the automotive industry due to their excellent electrical conductivity, thermal properties, corrosion resistance, and mechanical strength. Their applications span across both traditional internal combustion engine (ICE) vehicles and modern electric vehicles (EVs) (Table 6.11) [2, 3, 7, 15].

6.6 Summary

The automotive industry has become one of the important applications of copper alloy materials. With the development of automobiles towards electrification and intelligence, higher demands are placed on the complex performance characteristics of copper alloy materials.

Copper alloys play a key role in the automotive industry. They are a diverse range of metal compounds consisting primarily of copper and other elements such as zinc, aluminum, nickel, and tin. These alloys are characterized by exceptional mechanical properties, including high strength, corrosion resistance, and thermal conductivity, making them irreplaceable in the automotive industry.

Copper alloys for the automotive market include brass, bronze, and copper-nickel alloys that perform critical functions in manufacturing components such as heat sinks, connectors, terminals, and electrical systems. When comparing the various automotive copper alloy options, it is important to consider factors such as cost, availability, and specific application requirements. For example, although copper-nickel alloys have excellent corrosion resistance, they can be more expensive than other copper alloys. In such cases, brass or bronze alloys may be a viable alternative that provides satisfactory corrosion resistance at a lower cost. In addition, the availability of specific alloy compositions and their compatibility with manufacturing processes should be considered.

Table 6.11 Application of copper and its alloys in the automotive industry

Car system	Description
Electrical systems	
Wiring and connectors	Copper is widely used in vehicle wiring harnesses, connectors, and electrical terminals due to its superior conductivity
Electric motors	In electric and hybrid vehicles, copper windings are essential in motors and alternators
Batteries	Copper foils and busbars are used in lithium-ion and lead-acid batteries for efficient energy transfer
Sensors and electronics	Copper alloys are used in various sensors, relays, and electronic modules
Thermal management	
Radiators and heat exchangers	Copper-brass radiators provide efficient cooling solutions, especially in heavy-duty vehicles
Heat sinks	Used in electronic components and power modules to dissipate heat
Braking and friction components	
Brake pads and linings	Copper-based materials enhance friction performance and heat dissipation
Bushings and bearings	Copper alloys like bronze are used in bearings and bushings due to their wear resistance
Structural and mechanical components	
Gears and synchronizers	Copper alloys like brass are used in transmission components for smooth operation and durability
Fasteners and fittings	Brass and bronze fasteners resist corrosion and provide mechanical stability
Hydrogen and alternative fuel systems	
Fuel lines and injectors	Copper-nickel alloys are used in fuel systems due to their corrosion resistance and durability
Hydrogen fuel cells	Copper components play a role in hydrogen storage and energy conversion technologies
Weight reduction and sustainability	
Lighter components	Advanced copper alloys contribute to weight reduction in EVs and hybrid vehicles, improving efficiency
Recyclability	Copper is highly recyclable, making it a sustainable choice for automotive manufacturing

References

1. Jones B, Nguyen-Tien V, Elliott RJR (2023) The electric vehicle revolution: critical material supply chains, trade and development. World Econ 46:1–16. https://doi.org/10.1111/twec.13345

2. Bennett C (2022) Copper: the material of choice for vehicle manufacturers. https://copperalliance.org/resource/copper-the-material-of-choice-for-vehicle-manufacturers/. Accessed 25 Feb 2025
3. There are so many copper materials used in new energy vehicles! https://jtcopper.com/there-are-so-many-copper-materials-used-in-new-energy-vehicles.html. Accessed 25 Feb 2025
4. Singh R (2020) 9 - Nonferrous materials. In: Singh R (ed) Applied welding engineering, Third edn. Butterworth-Heinemann, UK, pp 85–91. https://doi.org/10.1016/B978-0-12-821348-3.00008-2
5. UNS standard designations for wrought and cast copper and copper alloys: introduction. https://www.copper.org/resources/standards/uns-standard-designations.html. Accessed 25 Feb 2025
6. MSP series (Cu-Mg alloy). https://www.mitsubishi-copper.com/en/products/materials/msp/. Accessed 27 Feb 2025
7. Copper alloys: the versatile wonders of metal engineering (2024). https://fastercapital.com/content/Copper-alloys--The-versatile-wonders-of-metal-engineering.html. Accessed 27 Feb 2025
8. Welter J-M (2014) Leaded copper alloys for automotive applications: a scrutiny. European Copper Institute, Bruxelles. https://elv.exemptions.oeko.info/fileadmin/user_upload/Consultation_2014_1/Ex_3/E3__02__Welter_2014_leaded_copper_alloys_for_automotive_applications-a_scrutiny.pdf. Accessed 27 Feb 2025
9. Ke J, Xie Z, Liu R, Jing K, Cheng X, Wang H, Wang X, Wu X, Fang Q, Liu C (2022) Development of Y_2O_3 dispersion-strengthened copper alloy by sol-gel method. Materials 15:2416. https://doi.org/10.3390/ma15072416
10. Carro G, Muñoz A, Savoini B, Monge MA, Pareja R (2019) Processing, microstructure and mechanical characterization of dispersion strengthened Cu-1%Y. Fusion Eng Des 138:321–331. https://doi.org/10.1016/j.fusengdes.2018.11.058
11. Zhang W-J, Huang L, Mi X-J, Xie H-F, Feng X, Ahn JH (2024) Researches for higher electrical conductivity copper-based materials. CMat 1:e13. https://doi.org/10.1002/cmt2.13
12. Yang Y, Kuang G, Li R (2022) Optimizing the electrical and mechanical properties of Cu-Cr alloys by Hf microalloying. Metals 12:485. https://doi.org/10.3390/met12030485
13. Varol T, Güler O, Akçay SB, Çolak H (2022) The evolution of microstructure and properties of Cu-Cr alloys synthesized via flake powder metallurgy assisted by mechanical alloying and hot pressing. Mater Today Commun 33:104452. https://doi.org/10.1016/j.mtcomm.2022.104452
14. Liao W, Hu Y, Liu Q (2022) High strength and high electrical conductivity C70250 copper alloy with fibrous structure reinforced by high density, multi-scale nano-precipitates and dislocation. Mater Sci Eng A 846:143283. https://doi.org/10.1016/j.msea.2022.143283
15. AngelinThangakani J, Dorothy Sheela C, Dorothy R, Renugadevi N, Jeyasundari J, Rajendran S, Behera A (2022) Applications of copper alloy nanoparticles in automotive industry (Chap. 14). In: Song H, Nguyen TA, Yasin G, Singh NB, Gupta RK (eds) In micro and nano technologies, nanotechnology in the automotive industry. Elsevier, pp 269–285. https://doi.org/10.1016/B978-0-323-90524-4.00014-1

Chapter 7
Titanium and Its Alloys

Abstract This chapter explores the properties, classifications, and applications of titanium and its alloys, emphasizing their growing role in the automotive industry. Titanium's high specific strength, excellent corrosion resistance, and heat tolerance make it ideal for engine components such as valves, rods, and mufflers. The chapter also highlights the mechanical properties influenced by microstructures and discusses surface treatments, particularly oxidation, to overcome poor wear resistance. Although initially used in high-performance vehicles due to high cost, recent technological advances aim to enable broader use in mass-produced cars to improve fuel efficiency and reduce CO_2 emissions. Tables and figures illustrate alloy compositions, historical usage, and property comparisons.

7.1 Introduction

Titanium has a density of 4.51 g/cm^3, between aluminum (2.7 g/cm^3) and iron (7.6 g/cm^3). Titanium alloy is a new structural and functional material with excellent overall properties, low density, and high specific strength [1]. The specific strength of titanium alloy is higher than that of aluminum alloy and steel, and the strength is close to that of steel. Titanium alloy has excellent corrosion resistance compared to stainless steel, especially in marine atmospheric environments, chloride ion erosion, and micro-oxidation atmospheres. It also maintains good ductility, and the service temperature of refractory titanium alloy can reach around 550 °C, making it significantly more heat resistant than aluminum and magnesium alloy. It also has good machinability and weldability [2]. Titanium is primarily used in internal combustion engine components such as valves, valve springs, holders, and connecting rods.

The use of titanium materials in the automotive industry began with Formula 1 racing cars in the 1980s, with the application primarily in engine parts [3] (Table 7.1).

Due to its excellent corrosion resistance, lightness, fuel efficiency, performance, and high strength-to-density ratio, titanium and its alloys have become widely used. However, due to their high price, their use in the automotive industry is limited, except for racing vehicles and vehicles for special purposes.

K. Berladir et al., *Materials for the Automotive Industry*, Springer Tracts in Mechanical Engineering, https://doi.org/10.1007/978-3-032-05866-9_7

Table 7.1 History of application of titanium alloys

Year	Manufacturer	Application
1989	Mitsubishi	Ti-22 V-4Al in the AMG engine retainers of the Gallant 1
1990	Honda motors	Ti-3Al-2.5 V + REM in the connecting rods of the sports cars NSX
1998	Toyota	Sintered titanium alloys Ti-6Al-4 V/TiB and Ti-Al-Zr-Sn-Nb-Mo-Si/TiB in the intake and exhaust engine valves, respectively, in the Altezza
1998	Kawasaki	Titanium alloys in the muffler of the large sports-type motorcycle ZX-9
2000	Nissan motor	Ti-6Al-4 V and Ti-6Al-2Sn-4Zr-2Mo-Si in engine inlet and exhaust valves, respectively, for the CIMA
2001	Volkswagen	Ti-4.5Fe-6.8Mo-1.5Al in suspension spring of Lupo FS
2001	General Motors	Titanium alloys in dual mufflers of the Corvette Z06

7.2 Classification of Titanium Alloys

Titanium is available in pure form and the form of alloys. These alloys are generally divided into three categories: alpha (α), alpha–beta (α + β) and beta (β) [4] (Table 7.2).

Commercially pure titanium (CP-Ti) contains elemental titanium plus some impurities such as N, H, O, and Fe. CP-Ti materials mainly contain α-phase and are divided into many classes. Strength increases with increasing amounts of O and Fe elements.

α-alloys are formed by one solid solution of the α-phase. These alloys have good properties at high temperatures but do not allow microstructural changes by heat treatment to improve their properties. They are mainly used in the chemical and processing industries, where excellent corrosion properties and deformability are of primary importance, with a high strength-to-weight ratio in second place.

Near-α alloys contain α-phase and usually less than 10% β-phase, obtained by adding a small amount (1–2%) of β-stabilizers. The β-phase improves the strength and machinability of alloys and helps achieve a balance between the higher strength of α + β alloys and the creep resistance of α-alloys. Today, their upper operating temperature is limited to around 500–550 °C. In this class, the most widely used commercial high-temperature titanium alloys are for aircraft engine applications.

The α + β alloys contain from 4 to 16% β-stabilizers. The most commonly used alloy in this category is Ti-6Al-4 V, which makes up more than 50% of all titanium alloys today. Commercial α + β alloys typically contain 10–20% β-phase at room temperature.

Near-β alloys, also called metastable alloys, contain small amounts of α-stabilizers and 10–15% β-stabilizers. Accordingly, the β-stabilizers help to preserve the β-phases under metastable conditions at room temperature, and these alloys can be heat-treated by aging, leading to the precipitation of a very fine α-phase dispersed in the β-matrix. In addition, these alloys optimize high strength and toughness and can be hardened to strength levels of more than 1400 MPa.

Table 7.2 Categories of titanium materials and selected types [5]

Category	Selected materials			
CP-Ti	CP-Ti (0.2Fe, 0.18O)	CP-Ti (0.3Fe, 0.25O)	CP-Ti (0.3Fe, 0.35O)	CP-Ti (0.5Fe, 0.40O)
α-alloy	Ti-5Al-2.5Sn	Ti-3Al-2.5 V	Ti-2Cu	Ti-0.3Mo-0.8N
Near-α alloy	Ti-5Al-6Sn-2Zr-1Mo-0.2Si	Ti-2Al-2Sn-4Zr-2Mo	Ti-8Al-1Mo-1 V	–
α + β alloy	Ti-6Al-4 V	Ti-6Al-4 V ELI	Ti-6Al-4 V-2Sn	Ti-6Al-2Sn-4Zr-6Mo
β-alloy	Ti-13 V-11Cr-3Al	Ti-11.5Mo-6Zr-4.5Sn	Ti-13 V-11Cr-3Al	–
Near-β alloy	Ti-6Al-2Sn-4Zr-6Mo	Ti-8Mo-8 V-2Fe-3Al	Ti-10 V-2Fe-3Al	Ti-8Mn
Ti-aluminides	Ti-24Al-10Nb	Ti-25Al-17Nb-1Mo	Ti-22Al-27Nb	–

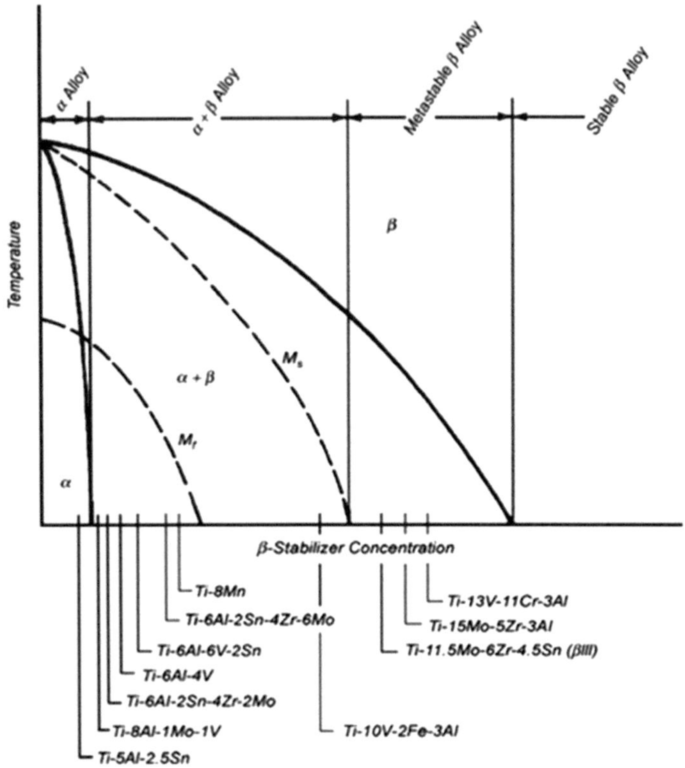

Fig. 7.1 Ti alloy compositions related to the phase diagram of the pseudo binary β isomorphous [6]

Titanium aluminides are a special class of alloys with exceptional physical and mechanical properties. These materials based on intermetallic compounds Ti_3Al (α2) and TiAl (γ) can increase operating temperatures to approximately 650 and 800 °C. Since Ti_3Al alloys have medium long-term stability, current research and development of titanium aluminide is focused on TiAl-based alloys.

Figure 7.1 summarizes the categories of typical titanium alloys, and Table 7.3 presents their main characteristics.

7.3 Titanium Alloys

There are six types of pure titanium (grades 1, 2, 3, 4, 7, and 11) and 4 types of titanium alloys (Ti 6AL-4 V, Ti 6AL ELI, Ti 3Al 2.5, Ti 5Al-2.5Sn) (Table 7.4) [7]. Titanium alloys typically contain Al, Mo, V, Nb, Ta, Zr, Mn, Fe, Cr, Co, Ni, and Cu [8].

Table 7.3 Categories and characteristics of the typical titanium alloys [1]

Categories	Typical alloys	Characteristics
α	Ti-5Al-2.5SnTi-6Al-2Sn-4Zr 2Mo	The general strength and toughness, good welding performance. Strong anti-oxidant, higher creep strength. Less used in the manufacture of golf masthead
α + β	Ti-6A1-4VTi-6Al-2Sn-4Zr 6Mo	The strength and toughness on the heat treatment can be strong, and with an excellent weld fatigue performance, multi-applications such as iron casting masthead, fairway
β	Ti-13 V-11Cr-3AlSp700Ti 15Ta-3Cr-3Al-3Ni	High strength, heat treatment, malleable and good cold forming performance, applicable to a variety of welding methods
TixAl	Ti3Al(a2)&TiAl	Desire to achieve 900-degree temperature, but poor toughness and plastic property at room temperature

Table 7.4 Main types of titanium alloy and their application [7]

Type of titanium alloy	Typical application
Ti 6AL-4 V (grade 5)	Aircraft turbines, engine components, aircraft structural components, aerospace fasteners, high-performance automatic parts, marine applications, and sports equipment
Ti 6AL-4 V ELI (grade 23)	Orthopedic pins and screws, orthopedic cables, ligature clips, surgical staples, springs, orthodontic appliances, joint replacements, cryogenic vessels, and bone fixation devices
Ti 3Al 2.5 (grade 12)	Shell and heat exchangers, hydrometallurgical applications, elevated temperature chemical manufacturing, marine and airfare components
Ti 5Al-25.Sn	Aircraft and airframe applications

Ti 6Al-4 V (Grade 5) is the most widely used of the titanium alloys and is referred to as the «workhorse» titanium alloy. It is believed to account for half of the world's titanium use. Ti 6Al-4 V can be heat treated to increase its strength. It can be used in welded structures at operating temperatures up to 600 °C. This alloy provides high strength at low weight, useful formability, and high corrosion resistance. These desirable properties make Ti-6AL-4 V a popular choice in several industries.

Ti 6AL-4 V ELI (Grade 23) is commonly referred to as surgical titanium due to its use in surgery. It is a purer version of grade 5 titanium alloy (Ti 6AL-4 V). It can be easily shaped and cut into small threads, coils, and wires. It has the same strength and high corrosion resistance as Ti 6AL-4 V. It is also light and highly resistant to damage by other alloys. Its use is highly desirable in the medical and dental field in complex surgical procedures because it has excellent biocompatibility, which allows it to be easily transplanted and attached to bone while being accepted by the human body.

Ti 3Al 2.5 (Grade 12) is a titanium alloy with better weldability. At high temperatures, it is as strong as other titanium alloys. This grade 12 titanium alloy is unique in that it exhibits the properties of stainless steel (one of the other strong metals). For example, it is heavier than other titanium alloys. Ti 3 Al 2.5 is most often used in the manufacturing industry, especially in engineering. It is highly resistant to corrosion and can be hot or cold formed.

Ti 5Al-2.5Sn (Grade 6) is a non-heat-treatable alloy that can achieve good weldability with stability. It also has high-temperature stability, high strength, and good corrosion resistance. It has a uniquely high resistance to creep (plastic stress over a long period of time, usually caused by extreme temperatures).

7.4 Properties of Titanium Alloys

The mechanical properties of titanium alloys depend on the individual properties of the α and β phases, their location, and volume fraction. α-phase has a lower density than the β-phase because the predominant element Al in the α-phase has a lower density than the predominant elements Mo or V in the β-phase [9–11].

In addition, fine microstructures increase strength and ductility, slow down crack initiation, and are a prerequisite for superplastic deformation. In contrast, coarse microstructures are more resistant to creep and fatigue crack growth. Equiaxial microstructures often have high ductility and fatigue strength. They are ideal for superplastic deformation, while lamellar structures have high fracture toughness and exhibit excellent resistance to creep and fatigue crack growth. Bimodal microstructures combine the advantages of plate and equiaxed structures, resulting in well-balanced property profiles.

Table 7.5 lists the properties of the most commonly used titanium alloys.

7.5 Application

In recent years, however, titanium and its alloys have come to be actively used for various parts of the general mass-produced cars (Fig. 7.2) due to the following factors [11]:

1. The demand for lightweight parts has become increasingly strict to prevent global warming by reducing CO_2 emissions.
2. Remarkable progress has been made in the development of technology for the manufacture of low-cost titanium.

The development and application of typical titanium components in the automotive industry are shown in Fig. 7.3.

Table 7.5 Properties of titanium alloys [5]

Category	Chemical composition (wt. %)	Hardness (HV)	Yield strength (MPa)	Tensile strength (MPa)	Elastic modulus (GPa)	Beta transus temperature T_{β} (°C)
α						
High purity Ti	99.98 Ti	100	140	235	100–145	882
CP-Ti grade 1	0.2Fe-0.18O	120	170–310	> 240	–	890
CP-Ti grade 4	0.5Fe-0.40O	260	480–655	> 550	100–120	950
Alloy grade 6	Ti-5Al-2.5Sn	300	827	861	109	1040
Near-α						
Ti-6-2-4-2-S	Ti-6Al-2Sn-4Zr-2Mo-0.1Si	340	990	1010	114	995
TIMETAL1100	Ti-6Al-2.7Sn-4Zr-0.4Mo-0.4Si	–	900–950	1010–1050	112	1010
TIMTETAL 685	Ti-6Al-5Zr-0.5Mo-0.25Si	–	850–910	990–1020	120	1020
α + β						
Ti-6-4	Ti-6Al-4 V	300–400	800–1100	900–1200	110–140	995
Ti-6-6-2	Ti-6Al-6 V-2Sn	300–400	950–1050	1000–1100	110–117	945
Ti-6-2-4-6	Ti-6Al-2Sn-4Zr-6Mo	300–400	1000–1100	1100–1200	114	940
Ti-17	Ti-5Al-2Sn-2Zr-4Mo-4Cr	400	1050	1100–1250	112	890
Near-β						
SP 700	Ti-4.5Al-3 V-2Mo-2Fe	300–500	900	960	110	900
Beta III	Ti-11.5Mo-6Zr-4.5Sn	250–450	800–1200	900–1300	83–103	760
Beta C	Ti-3Al-8 V-6Cr-4Mo-4Zr	300–450	800–1200	900–1300	86–115	795
Ti-10–2-3	Ti-10 V-2Fe-3Al	300–470	1000–1200	1000–1400	110	800
Ti-15–3	Ti-15 V-3Cr-3Al-3Sn	300–450	800–1000	800–1100	80–100	760

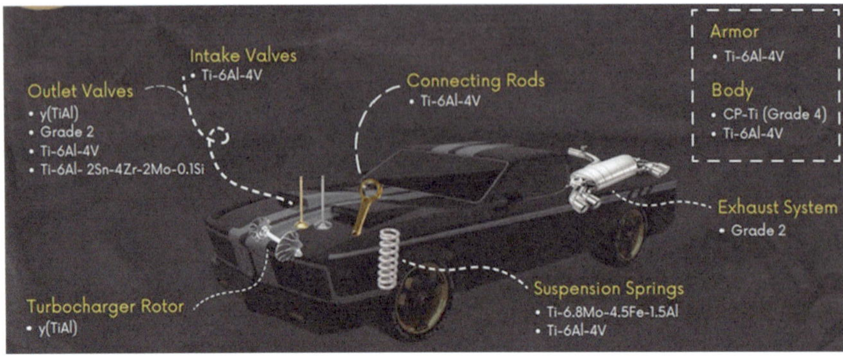

Fig. 7.2 The application of titanium and its alloys for various parts of the general mass-produced cars

The most difficult problem in using titanium alloys for intake valves is the development of wear-resistant surface treatment technology because it is generally known that titanium alloys have poor wear resistance. Surface treatments, including TiN hard coating, Mo thermal spraying, and Cr electroplating, are often used [11].

However, all these methods are expensive and unsuitable for maintaining long-term wear resistance. «Oxidation treatment» can use solid solution oxygen hardening, characterized by forming a relatively thick hardened layer in which highly concentrated oxygen diffuses into the surface layer of titanium to increase hardness.

Oxidation treatment is essentially heat treatment to heat to high temperatures in air. This leads to adopting a needle-like structure, which is superior in creep resistance, instead of equiaxed or mill-annealed microstructures, which are inferior in creep resistance.

7.6 Summary

Titanium and its alloys have gained significant importance in the automotive industry due to their exceptional strength-to-weight ratio, corrosion resistance, and heat resistance. Initially introduced in Formula 1 racing cars in the 1980s, titanium components such as intake and exhaust valves, connecting rods, and mufflers have since been incorporated into high-performance and specialized vehicles. However, the widespread adoption of titanium in mass-produced automobiles has been limited due to its high cost, despite ongoing efforts to develop low-cost manufacturing technologies.

Titanium alloys are classified into different categories, including α, α + β, and β alloys, each offering unique properties and applications. α-alloys, primarily used in chemical industries, exhibit excellent corrosion resistance, while α + β alloys, particularly Ti-6Al-4 V, dominate aerospace and automotive applications due to their balanced strength and ductility. β-alloys, known for their high strength and

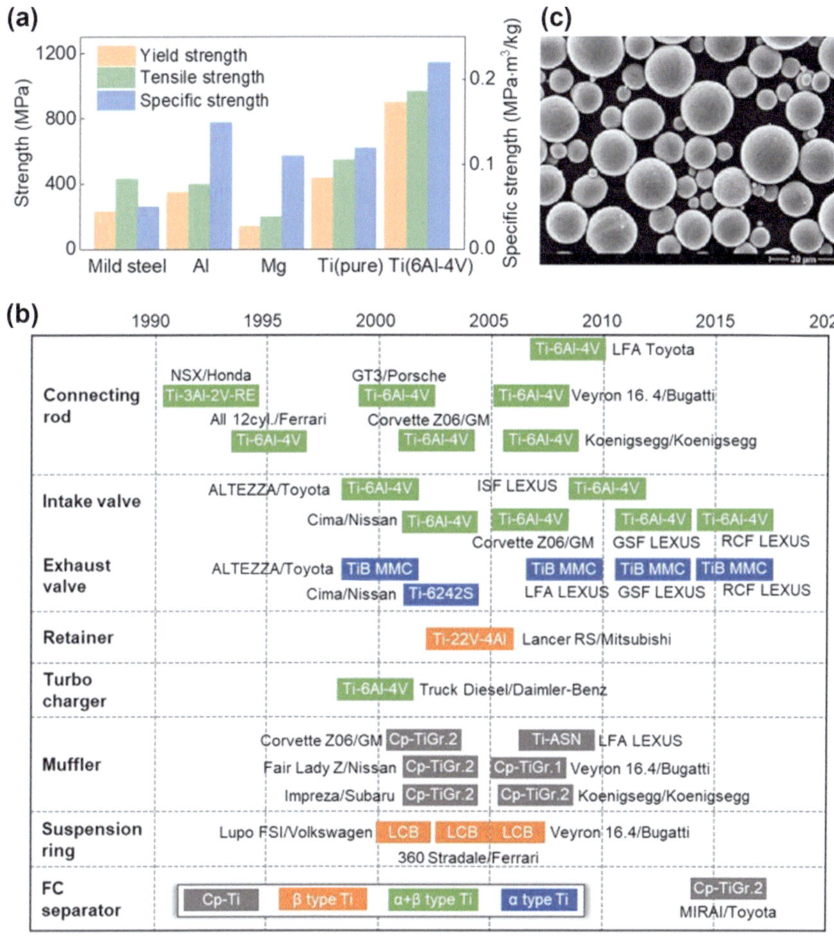

Fig. 7.3 **a** Comparison of mechanical properties between various types of metallic materials, including mild steel, aluminum, magnesium, and Titanium (pure and 6Al-4 V); **b** chronological development of representative titanium components in the automotive industry; **c** titanium powder processed by additive manufacturing [12]

cold-forming capabilities, are increasingly used in structural components. Additionally, titanium aluminides (Ti3Al and TiAl) are being developed for high-temperature applications, particularly in aerospace engines, though their limited toughness remains a challenge.

The mechanical properties of titanium alloys depend on their microstructure, with fine-grained structures enhancing strength and ductility, while lamellar structures provide better resistance to creeping and fatigue. Recent advancements in titanium applications in mass-produced vehicles focus on reducing weight to enhance fuel efficiency and lower CO_2 emissions. However, challenges such as the material's high cost and poor wear resistance, particularly in engine components, must be addressed.

Surface treatment technologies, such as oxidation treatment, are being explored to enhance wear resistance, and further collaboration among material producers, part manufacturers, and automakers is necessary to expand titanium's role in the industry.

References

1. Cui C, Hu B, Zhao L, Liu S (2011) Titanium alloy production technology, market prospects and industry development. Mater Des 32(3):1684–1691. https://doi.org/10.1016/j.matdes.2010.09.011
2. Besisa NHA, Yajima T (2024) Titanium-Based Alloys: Classification and Diverse Applications. IntechOpen, UK. https://doi.org/10.5772/intechopen.1005269
3. Yamashita Y, Takayama I, Fujii H, Yamazaki T (2002) Applications and features of titanium for automotive industry. Nippon steel technical report 85, pp 11−14. https://www.nipponsteel.com/en/tech/report/nsc/pdf/8503.pdf. Accessed 28 Feb 2025
4. Pushp P, Dasharath S M, Arati C (2022) Classification and applications of titanium and its alloys. Mater Today Proc 54(Part 2):537−542. https://doi.org/10.1016/j.matpr.2022.01.008
5. Veiga C, Davim JP, Loureiro AJR (2012) Properties and applications of titanium alloys: a brief review. Rev Adv Mater Sci 32:14–34
6. Najafizadeh M, Yazdi S, Bozorg M, Ghasempour-Mouziraji M, Hosseinzadeh M, Zarrabian M, Cavaliere P (2024) Classification and applications of titanium and its alloys: a review. J Alloys Compd Commun 3:100019. https://doi.org/10.1016/j.jacomc.2024.100019
7. Sachdev AK, Kulkarni K, Fang ZZ et al (2012) Titanium for automotive applications: challenges and opportunities in materials and processing. JOM 64:553–565. https://doi.org/10.1007/s11837-012-0310-8
8. Boyer RR (2010) Attributes, characteristics, and applications of titanium and its alloys. JOM 62:21–24. https://doi.org/10.1007/s11837-010-0071-1
9. Salihu SA, Suleiman YI, Eyinavi AI (2019) Classification, properties and applications of titanium and its alloys used in automotive industry-a review. Am J Eng Res 8(8):92–98
10. Takahashi K, Mori K, Takebe H (2020) Application of titanium and its alloys for automobile parts. MATEC Web Conf. 321:02003. https://doi.org/10.1051/matecconf/202032102003
11. Application of titanium alloy in automobile. https://titanium.net/application-of-titanium-alloy-in-automobile/. Accessed 28 Feb 2025
12. Zhang W, Xu J (2022) Advanced lightweight materials for automobiles: a review. Mater Des 221:110994. https://doi.org/10.1016/j.matdes.2022.110994

Chapter 8
Plastics

Abstract The chapter explores the extensive role of plastics in the automotive industry, highlighting their technical properties, historical development, and applications. Plastics, defined as organic polymers with additives, are valued for their light weight, corrosion resistance, thermal insulation, and moldability. Key categories include thermoplastics, thermosets, and elastomers. The chapter reviews specific polymers—such as polypropylene, polyurethane, PVC, ABS, polyamide, and PET—used in vehicle components, including bumpers, dashboards, and seats. Additionally, it examines innovations in bio-based plastics and the rising importance of recycled plastics, driven by environmental concerns. The chapter concludes by emphasizing plastics' irreplaceable role in modern car design, weight reduction, cost-efficiency, and sustainability, while addressing end-of-life recycling challenges and closed-loop systems.

8.1 Introduction

From a technical point of view, the term plastics encompasses a wide group of materials with specific properties, consisting of organic polymers or their mixtures with suitable additives. They are composed of macromolecules containing thousands of atoms of hydrogen, carbon, oxygen, and nitrogen (and possibly atoms of other elements).

The origins of plastics date back to the first half of the 19th century (Fig. 8.1).

Among the characteristic properties of plastics are [1]:

- low weight,
- low density (most plastics have a density between 1.1 and 1.6 g/cm^3, for comparison: Mg → 1.75 g/cm^3),
- corrosion resistance (as well as to other chemical formulas),
- low thermal conductivity,

K. Berladir et al., *Materials for the Automotive Industry*, Springer Tracts in Mechanical Engineering, https://doi.org/10.1007/978-3-032-05866-9_8

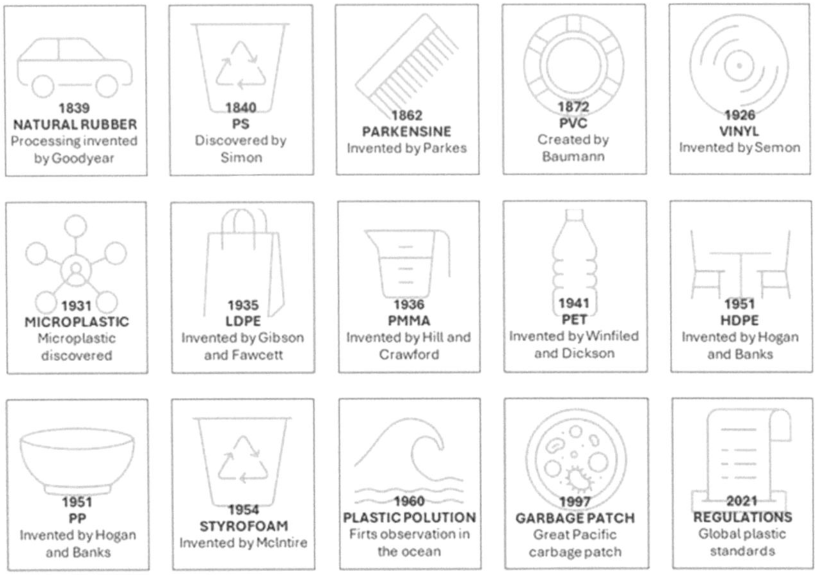

Fig. 8.1 The development of plastics over time

- easy to mold and work due to their elasticity (type of manufacturing processes: blow/injection/rotational molding, extrusion, casting, vacuum forming, 3D printing, and CNC machining),
- color variability (using pigments) and design flexibility,
- affordability.

On the other hand, plastics do not excel in mechanical properties—compared to metals (e.g., they have low impact toughness values—except for ABS, HDPE, PC, etc.).

The classification of plastics according to their behavior during shaping—at higher temperatures:

- thermoplastics—can be reversibly melted by heating and resolidified by cooling without significant changes of mechanical and optical properties (consist of macromolecular chains with no cross-links between the chains),
- thermosets—heat curable (narrow cross-linked molecule chains), and elastomers (plastics with wide netlike crosslinking between molecules, they cannot be melted without degradation of the molecule structure—usually) (Fig. 8.2).

The proper selection of plastic is determined by the conditions of use. Therefore, before using any plastic, specific framework conditions—purpose and applications—must be known. Based on this information, qualified experts can compare the requirements of technical values and perform an assessment for material selection. Currently, there is an increasing demand for the utility properties of plastic products, as well as for reducing production costs [1].

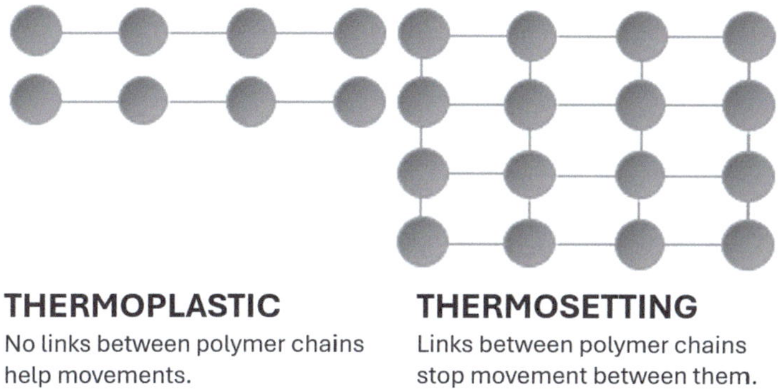

Fig. 8.2 Thermoplastic versus thermosetting—differences

The plastics industry is a key enabler of innovation of many products and technologies in other sectors of the economy like healthcare, energy generation, aerospace, automotive, maritime, construction, electronics, packaging or textile. None of these sectors would innovate and grow as much as they do without plastic materials and solutions. In Europe, packaging applications are the largest application sector for the plastics industry and represent 39.6% of the total plastics demand. Building and construction is the second largest application sector with 20.3% of the total European demand. Automotive is the third sector, with a share of 8.5% of the total demand (Fig. 8.3). Electrical and electronic applications represent 5.6% of the plastics demand and are closely followed by agricultural applications, which have a share of 4.3%. Other application sectors such as appliances, household and consumer products, furniture, and medical products comprise a total of 21.7% of the European plastics demand [2].

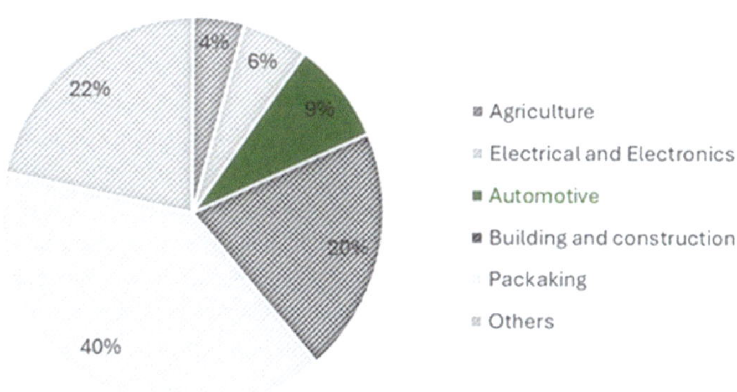

Fig. 8.3 Applications of plastics in sectors

Plastic is the second most commonly used material in vehicles (after metals and alloys). Modern cars contain several plastic parts and components. The use of plastics in the car manufacturing industry presents several advantages, such as [3–7]:

- a vehicle weight reduction, which leads to lower fuel consumption and a decrease in emissions of Green House Gasses (GHG). For example—1 kg of vehicle saving the environment about 20 kilos of carbon dioxide. According to Martin Geode et al.: 100 kg mass reduction in a car weight could result in 0.3–0.5 dm^3/100 km;
- increasing passenger safety. For example—seatbelts (polyamide or polyester fibers), airbags (high-strength nylon or reinforced polyamide fibers), infant carriers/child seats/booster seats are crucial components for car safety (reducing the risk of deaths and injuries);
- cost-effective design opportunities—to create innovative design concepts on vehicles that enhance passenger comfort at a realistic cost;
- reducing vehicle costs and providing resource-efficient solutions;
- can also be recovered when a vehicle reaches its end-of-life stage (recycled automobile materials are addressed in a separate subchapter).

8.2 History of Automotive Plastics

Plastics have been with the automotive industry from the very beginning. The earliest thermoplastic material, celluloid, was invented in 1855 by Alexander Parkes, who named his invention *Parkesine.* And only thirty years later the first vehicle to be considered a car, the Carl Benz Patent-Motorwagen 1, was built. The first attempt to build a car made entirely of plastic was made by Henry Ford in 1941, due to steel rationing during the war. It was a *Soybean car* = … «soybean fiber in a phenolic resin with formaldehyde used in impregnation» → cutting weight to 907 → 454 kg lighter than a steel car. However, the mass use of plastics started with the production of the Ford T model in 1951 [8–10].

These materials (together with reinforced composite versions of some bulk plastics) gradually acquired respectability in the automotive industry as it came to be seen that they could be produced. The second big factor was the appearance, in this same period, of a seemingly endless flow of exciting new materials. Most materials, now classed as engineering plastics, were in volume production in their basic formulations before the end of the 1960s. The spectacular growth in the plastics market since the early 1970s has resulted not so much from the discovery of new plastics as from the refining and targeting of the established ones. From the beginning, much of the development work on engineering plastics was directed towards the motor industry. Plastics milestones in the automotive industry are presented in Table 8.1 [11].

Table 8.1 Plastics milestones in automotive—selected parts [11]

Date	Part	Kind of plastic	Model/Source
Pre-1950	Distributor, switches	PF	Chevrolet Corvette
	Steering wheel shroud	PF	
	Colored knobs and handles	CA, CAB	
	Leathercloth	PVC	
1950–1954	Cable sheathing	PVC	
	Car body (panels)	GRP	
	Battery case	PS	NSU Prinz
	Gear wheels	PA	
	Rear light lenses	PMMA	
1955–1959	Roof	GRP	Citron DS
	Truck cabs	GRP	
	Rotationally cast armrests	PVC	
	Vacuum-formed interior panels	ABS	
	Foam-backed fascia skin	PVC	Mercedes
	Trim fasteners	PA	
1960–1964	Gas pedal	PP	GM USA
	Injected molded cooling fan	PP/PA	Rootes, BL
	Improved distributor cap	GF Alkyd	
	Carpet backing	LDPE	
	Door/window component	PA/POM	
	Covers	PP/ABS	
	Kick plates	PP	Vauxhall
1965–1969	Injection molded heater case	PP	
	Seat foam	PU	
	Front grille	ABS	Renault 6
	Expansion tank	PP	
	Hydraulic fluid reservoirs	PP	
	Headlamp lenses	PC	
1970–1974	Battery box	PP	Lucas
	Bumpers (rigid)	SMC	Renault 5
	Lateral protection panels	SMC	Renault 5
	Fuel tank	HDPE	VW Passat
	Front and rear spoilers	PU	BMW
	Two-piece car body	GRP	Lotues Elite
	Headlamp body	DMC	Lucas for Vauxhall
1975–1979	Painted RTM panels	Epoxy	Matra Murena
	Door-mounted mirrors	GR-PA	Renault 14

(continued)

Table 8.1 (continued)

Date	Part	Kind of plastic	Model/Source
	Painted grille	SMC	Ford Cortina
	Flexible lateral protection strip	PU	Mercedes
	Flexible bumper covers	PP/EPDM	Fiat, Citroen
	Engine compartment under-panel	GTM	USA
	Oil pan	STX	GM USA
	Estate car seat-back/load floor	HDPE	USA
1980–1984	Bonnet and tailgate (volume prod.)	SMC/XMC	Citroen BX
	Self-supporting bumper	PC/PBT	Ford Sierra
	Painted thermoplastic panel	ABS, GR-PA	Ford Sierra
	Full-face wheel trim	GR-PA	Ford Sierra
	Slush-molded fascia	PVC	Audi Quattro
	Bumper beam	SMC	Ford Australia
	Car body (panels. volume prod.)	Various	Pontiac Fiero
	Painted PP bumpers	PP	Rover 2(X)
	Painted thermoplastic body panels	PC/ABS	Honda CR-X
	Steering wheel + integral skin foam	PU	USA
1985–	On-line painted TP body panels	PPE/PA	Nissan Be-l
	Painted KTM panels (volume prod.)	Epoxy	Renault Espace
	Inlet manifolds (fusible core)	BMC/GR-PA	Ford. BMW
	Body platform	GR-PU/Epoxy	BMW ZI
	Injection molded under panel	GR-PP	Volvo 400
	Composite drive shaft	CR + CR epoxy	Renault Espace (Quadra)

8.3 Kinds of Plastics Used in the Automotive Industry

The average modern car weighting 1500 kg contains between 12 and 15% of plastic materials. This equates to over 2000 plastic parts of all shapes and sizes. With the dynamic progress in plastics processing technologies, the range of applications in modern vehicle construction continues to grow (Fig. 8.4).

Designers and constructors have more and more opportunities thanks to the excellent quality and optimum parameters of plastic parts (Fig. 8.5) [7, 8].

Describing some kinds of plastics that are most used in the automotive industry [6, 12–16].

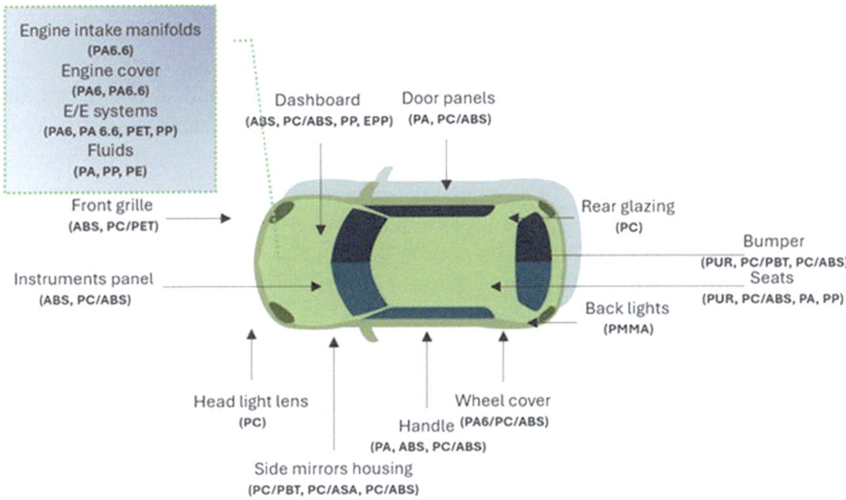

Fig. 8.4 Plastics application in automotive parts

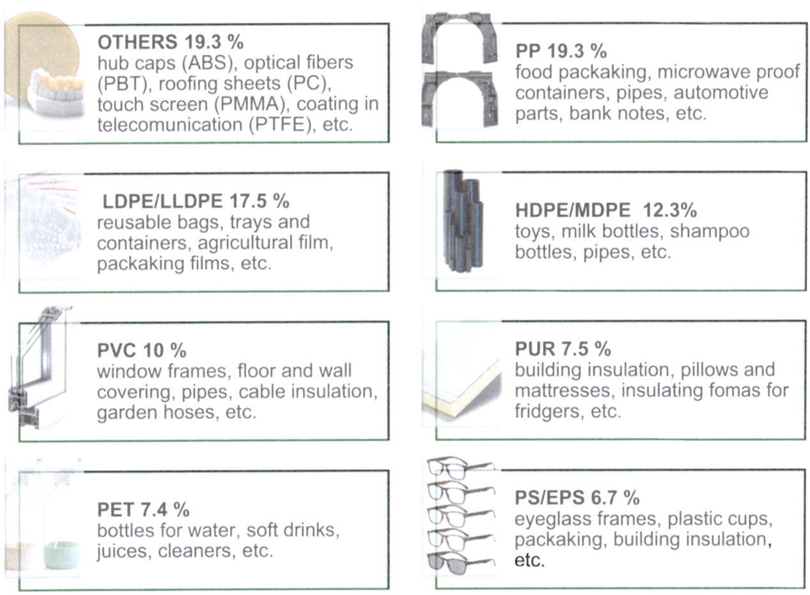

Fig. 8.5 European plastics converter demand by polymer types in 2016

8.3.1 Polypropylene

Polypropylene (PP) is low in cost but has outstanding mechanical properties and moldability; it accounts for more than half of all the plastic materials used in automobiles. PP belongs to the group of polyolefins. It is produced in the process of a monomer (propylene) additional polymerization. Polymerization results in polypropylene—a material with high strength, toughness, and resistance to corrosion, abrasion, impact, acids, and bases. The chemical and electrical properties are similar to polyethylene, but it has better mechanical properties (tensile strength of 30–35 MPa) and is relatively lightweight (0.90–0.91 g cm^{-3}). Polypropylene has lower water absorption than polyethylene, 0.01% at 24 h submersion (some studies report a value of 0.008% at 24 h submersion). As for the thermal expansion of polypropylene, it is similar to that of HDPE, PP, PVC, ABS, and nylon (Fig. 8.6).

Black PP has the best UV resistance and is increasingly used in the construction industry. PP is a widely applicable plastic in cars due to its ability to adapt to many plastic fabrication techniques.

Therefore, it is critical to make the following automotive parts and components:

- bumper fascia,
- gas cans,
- engine covers,
- cable insulation,
- instrument panels,
- replace hard rubber for battery housing.

To improve the performance of PP compounds are added inorganic fillers (Table 8.2, Fig. 8.7).

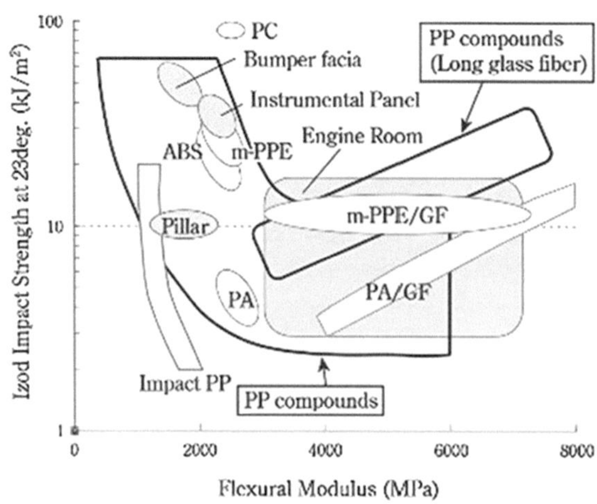

Fig. 8.6 Mechanical properties of PP compounds [15]

Table 8.2 Organic and inorganic fillers for PP compounds [15]

Inorganic compound	Oxide	Silica, titanium oxide, magnesium oxide, antimony oxide
	Hydroxide	Aluminum hydroxide, magnesium hydroxide, calcium hydroxide
	Carbonate	Calcium carbonate, Dolomite
	Sulfate	Basic magnesium SULFATE
	Silicate	Talcum, clay, mica, glass fiber/balloon/beads, calcium silicate, bentonite
	Carbon	Carbon black, graphite, carbon fiber
Organic compound	Powder/fiber	Wood powder, natural fibers from plants, aromatic polyamide fiber, etc.

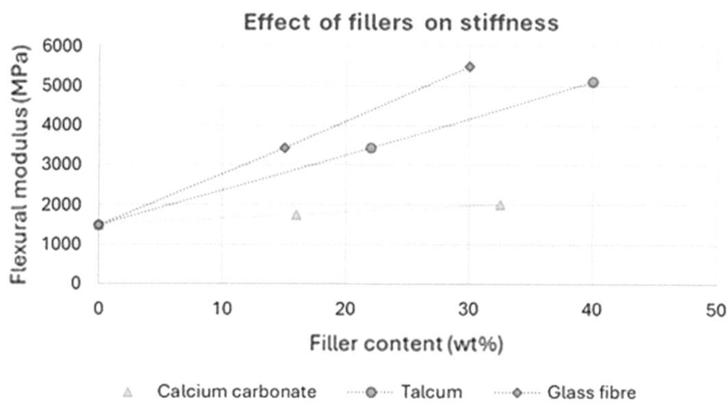

Fig. 8.7 Effect of fillers on stiffness

8.3.2 Polyurethane

Polyurethane (PUR) is highly suitable for the industry due to its insulative properties, strength, resilience, and moldability. Therefore, PUR is applicable in making parts and components that shield the noise and heat from car engines and parts that require strength in the car body. A high percentage of the total world PUR market is in foam—which results from the formation of gas bubbles in the polymerization process (Fig. 8.8).

There are three distinct types of PUR foam:

1. *low-density flexible foams*—excellent as resilient cushioning material, universally used in car seating. Semi-rigid variants have good shock-absorbing properties and are much used in automotive trim–filling for bumpers, backing material for the ABS/PVC–skinned fascia;
2. *low-density rigid foams*—excellent insulating materials, rarely applied in the automotive sector;

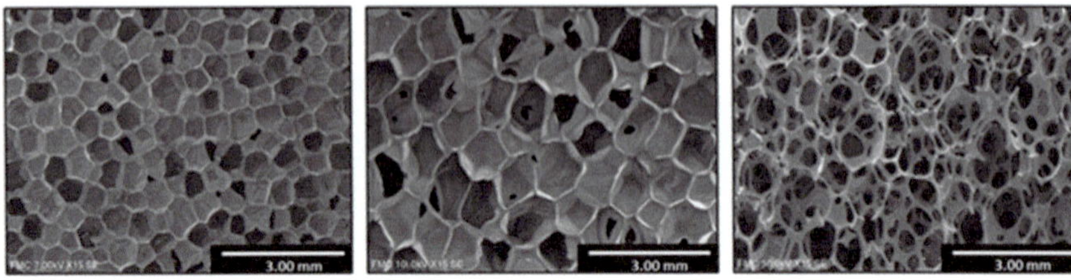

Fig. 8.8 Typical cellular structures for (from left side) closed cell rigid PU foam, hybrid rigid PU foam, and open cell PU foam [17]

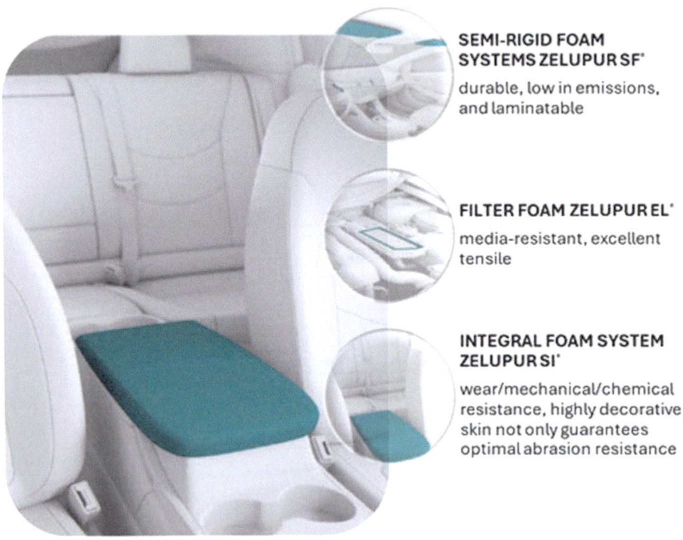

Fig. 8.9 Products of Zelu company, application of PUR foam

3. *high-density flexible foams*—foamed cellular core, but a relatively solid skin. Applicable for integrally skinned trim in the automotive industry.

The examples of PUR foam application are presented in Fig. 8.9.

8.3.3 Polyvinylchloride

Polyvinylchloride (PVC) has good resistance to chemical and solvent attacks. Its vinyl content gives it good tensile strength, and some grades are flexible. Colored or clear material is available. PVC has good electrical properties (425–1300 V/mil dielectric strength) and medium heat-deflection temperature. PVC homopolymer has

Fig. 8.10 PVC parts floor for HONDA

poor thermal stability and high brittleness (for example, compared to HDPE), further requiring additives to improve its properties.

Compounds added to PVC include plasticizers (to impart flexibility to polymeric materials), heat stabilizers (to prevent thermal stability during processing), lubricants (to aid the melt flow of PVC compounds during processing and prevent adhesion to metal parts), fillers (to lower the cost), and pigments. PVC has a higher water absorption than PE and PP—0.1% at 24-h submersion.

It is applicable in making the following components and parts (Fig. 8.10):

- vehicle airbags,
- door panels,
- cables,
- dashboards.

8.3.4 Acrylonitrile–Butadiene–Styrene

Acrylonitrile–butadiene–styrene (ABS) consists of discrete, crosslinked polybutadiene rubber particles that are grafted with poly(styrene-co-acrylonitrile) and embedded in the SAN matrix. It is a durable thermoplastic, resistant to weather/some chemicals and toughness, popular for vacuum formed components. It is a rigid plastic with rubber-like characteristics, which gives it good impact resistance. The polymer is strong, durable, and exhibits good electrical insulating properties.

ABS is applicable predominantly in interior use although there are also some exterior parts. Interior applications include instrument panels, knobs, and trim, while exterior parts include wheel covers, mirrors, and headlight housing.

Table 8.3 Characteristics of nylon 6/6 versus nylon 6 [18]

Nylon 66 has these advantages over nylon 6	Nylon 6 has these advantages over nylon 6/6
Higher temperature resistance	Slightly lower density
Higher strength	Better toughness
Higher stiffness	Better surface appearance
Lower moisture absorption	Lower processing temperature
Better abrasion resistance	–

8.3.5 *Polyamide*

Polyamide (PA, nylon 6/6, nylon 6) is a member of the engineering plastics family and offers superior load-bearing capability at elevated temperatures, good toughness, low-frictional characteristics, and good chemical resistance/high abrasion resistance. They also absorb water easily; it is necessary to use additives—to improve its flaws (for example, fiberglass to reduce water absorption capacity). Some types of nylons are produced by a stepwise polymerization of a dibasic organic acid with a diamine (there are many types of nylons, and the repeating unit is different for each type). For example, Nylon 6/6 is produced by a polymerization reaction between hexamethylene diamine and adipic acid to produce polyhexamethylene diamine.

The automobile sector is the largest consumer of PAs, accounting for 35% of PA consumption. Parts and components made with the polymer include (Table 8.3):

- engine covers,
- door handles,
- gear,
- fan blades,
- seat belts,
- car upholstery,
- fuel caps and lids.

8.3.6 *Polyethylene*

Polyethylene (PE) has good chemical and microbial resistance. Two types are used, low density polyethylene (LDPE) and high-density polyethylene (HDPE) can be manufactured in a range of densities. LDPE has a branched-chain structure. It was first commercially produced in the United Kingdom in 1939 by using autoclave reactors requiring pressures over 100 MPa and temperature of about 300 °C. In contrast, HDPE has essentially a straight-chain structure. HDPE was first produced commercially by the Phillips and Ziegler processes by using special catalysts in 1956–1957; in these processes—the pressure and temperature for the reaction to convert ethylene to polyethylene were considerably lowered.

Table 8.4 Characteristics of LDPE/HDPE [19]

LDPE	HDPE
More branching	Less branching, more linear
Flexible	Semi-rigid, tough
Good compression set	Good compression set
Good abrasion resistance	Excellent abrasion resistance
Lightweight	Lightweight
Good chemical resistance (variable)	Good chemical resistance (variable)
Weatherproof	Weatherproof
Low cost	Low cost

The fundamental differences between HDPE and LDPE are presented in Table 8.4. PE is suitable for making plastic fuel tanks and GR car bodies.

8.3.7 Polystyrene

Polystyrene (PS) is the fourth-largest tonnage thermoplastic. PS is a naturally transparent polymer that has solid and foamed forms. It is insoluble in water, easy to manufacture, has heat and chemical resistance, and has good optical properties.

The homopolymer is characterized by its rigidity, sparkling clarity, and brittle. But PS has poor resistance to UV light.

The commonly manufactured products in the automobile industry include bumper cores, boot in-fills, void fillers, roof liners, headrests, head impact, knee bolsters, side-impact protection, car seating, sun visors, car air conditioning liners, under bonnet battery liners, under bonnet sound deadening and material handling dunnage.

8.3.8 Polyoxymethylene

Polyoxymethylene (POM) has big stiffness, rigidity and excellent yield, which are stable in low temperatures as well. Very good chemical, UV, and fuel resistance. As a result, it is applicable in interior and exterior trims and fuel system parts.

8.3.9 Polycarbonate

Polycarbonate (PC) has good weather and UV resistance, with transparency levels almost as good as acrylic. It possesses characteristics such as superior toughness, high impact strength, rigidity, fire resistance, high dimensional stability, low moisture absorption, resistance to corona discharge, high heat deflection, and it is extremely lightweight. In comparison with PVC—it is less toxic, recyclable, and eco-friendly. They are applicable in making bumpers, bullet-proof glass, and headlamp lenses.

8.3.10 Polymethyl Methacrylate

Polymethyl methacrylate (PMMA) is an alternative to glass (more impact resistance than glass), has a reasonable tensile strength (shatterproof grades are available), and has good UV/weather resistance. The trade name of PMMA is Plexiglas. PMMA is cheap, available, and suitable for making the following parts:

- car light covers,
- windshields for motorcycles,
- automotive topcoats.

8.3.11 Polybutylene Terephthalate

Polybutylene terephthalate (PBT) has good chemical resistance, electrical properties, hard and tough material with low moisture-absorption characteristics, very good resistance to dynamic stress, thermal/dimension stability, and easy processability. However, compared to plastic polymers such as PET, it has lower strength/rigidity and better impact resistance. PBT applications:

- large exterior-body components,
- coil bobbins,
- high-energy ignition caps and rotors,
- plug connectors,
- bumpers,
- door handles,
- windshield wipers,
- door mirror system/headlamps,
- antilock braking systems.

8.3.12 Polyethylene Terephthalate

Polyethylene terephthalate (PET) has a glass appearance, good thermal stability/ electrical properties, high impact strength, lightweight, toughness, easy moldability, very low water absorption, excellent surface properties, and does not produce odors or fumes. The main application of PET is in drinks bottles and food containers. It is applicable in exterior body casing; engine covers and connector housing.

8.3.13 Acrylonitrile Styrene Acrylate

Acrylonitrile styrene acrylate (ASA) has great toughness and rigidity, good chemical resistance and thermal stability, outstanding resistance to weather, aging and yellowing, and high gloss. Application: for interior panels.

8.3.14 Styrene-Acrylonitrile

Styrene-acrylonitrile (SAN) resins have better chemical resistance, higher heat-deflection temperatures, higher toughness, better load-bearing characteristics than PS. SAN are rigid and hard. Major applications:

- automotive instrument lenses,
- dash components,
- glass-filled support panels.

The following graphical representations compare the properties/characteristics: tensile strength, compressive strength, IZOD impact strength, and specific weight of individual polymers (Fig. 8.11).

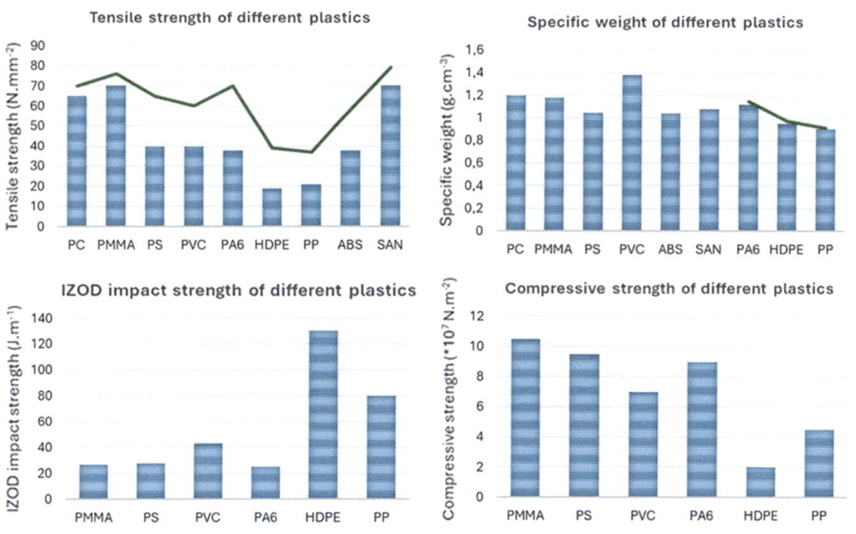

Fig. 8.11 Plastic property comparison graphs. Note: a bar chart with a curve shows the range of values of the displayed characteristic (for example, tensile strength for PC is in the range: 65–70 N mm^{-2})

8.4 Bio-based Plastics Used in Automotive Industry

The term *biopolymer* is a technical synonym of bio-based polymers (which may be or may not be biodegradable in nature). Application of biopolymers in manufacturing car body parts effectively improves fuel efficiency, design flexibility, comfort, insulation, and reduces our dependence on fossil fuel (open to new opportunities for a transition in the name of reducing the carbon footprint—CFP).

Bio-based plastics can divide into two groups:

1. Bio-based or partially bio-based, nonbiodegradable plastics such as [20]:
   a. mass commodity plastic (e. g. bio-based PE, or bio-based PET);
   b. bio-based technical performance polymers e. g. polytrimethylene terephthalate (PTT) or thermoplastic, copolyester elastomers (TPC-ET) or bio-based polyamides;
2. Plastics that are both bio-based and biodegradable e. g. PLA (polylactic acid), PHA (polyhydroxyalkanoates) and PBS (polybutylene succinates) but can also be used in durable applications.

Bio-based plastics used in automotive sector according to [21] including bio-based polyamides, PTT, and bio-based polyolefins (e. g. PLA-blends).

Figure 8.12 demonstrates the main differences between bioplastics and plastics.

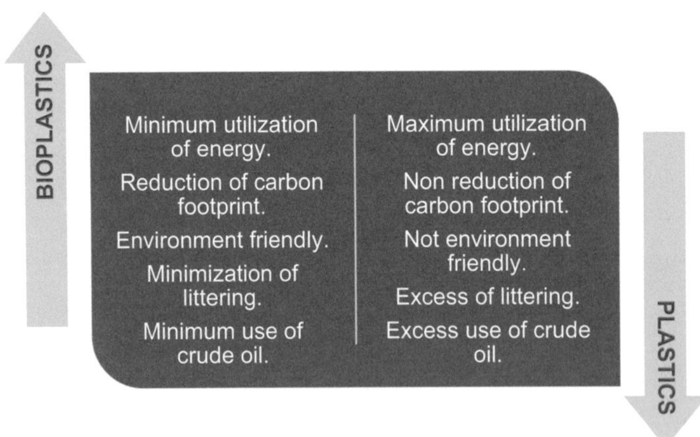

Fig. 8.12 Bioplastics versus petroleum-based plastics

8.4.1 Polylactic Acid

Polylactic acid (PLA) is a thermoplastic polyester derived from renewable resources such as corn starch, tapioca products, and sugarcanes (an interesting fact is only 1.6 kilos of biomass are required to make a kilo of PLA) (Fig. 8.13).

PLA blends—polymers made from PLA but blended with other polymers such as PP or PET. PLA is biodegradable and can be recycled (by broken down to its monomer by a thermal depolymerization process or by hydrolysis). PLA has better mechanical properties compared to commodity plastics such as PP in terms of tensile strength, elongation, and flexural strength (tensile strength = 59 MPa, elongation = 3%, flexural strength = 85 MPa, Izod impact strength = 33.4 J m^{-1}). The disadvantages of PLA becoming a commodity plastic are cost, poor thermal and impact properties, and hydrolytic instability. The production of PLA components is similar to other thermoplastic production for the automotive industry and is done by injection molding [22–25].

A study [23] states that a car part made of PLA can completely degrade in three to six months after its lifespan. Other bio-based plastics do not or only barely degrade. It should be mentioned that PLA needs a temperature of at least 60 °C, a humid environment, and a diet of digestive microbes. These conditions can only be achieved in an industrial compost factory. A one-way trip to a normal garbage dump is not an option. According to NatureWorks, PLA breaks down as slowly as a PET bottle under normal circumstances: it takes between 100 and 1 000 years.

The examples of PLA application in the automotive industry is presented in Table 8.5.

There are a number of methods for recycling PLA [22]:

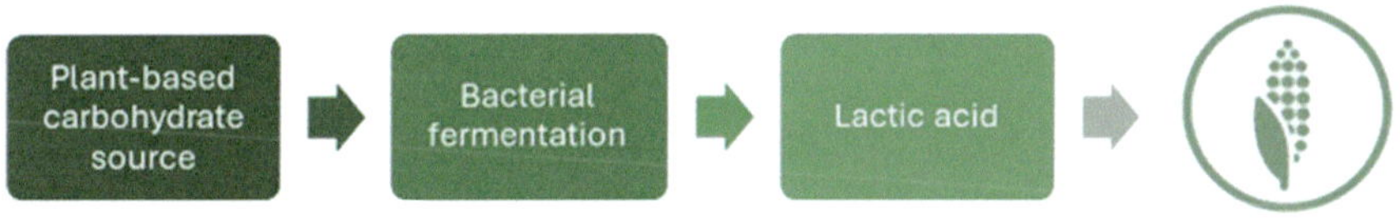

Fig. 8.13 PLA production process

Table 8.5 PLA for automotive application [26, 27]

Company	Brand	Model	Source	Material	Application
Mazda	Mazda	Mazda 5 Hydrogen RE Hybrid	Corn	Polylactic acid	Console, seat covers
Toyota	Lexus	ES300	Kenaf	Polylactic acid	Package shelves
Ford		U model	Wheat straw Coconut Guava plant	Polylactic acid	Roof, floor carpets

Fig. 8.14 Castor plant → castor bean → castor oil → polymerization → polyamides

- mechanical recycling,
- and processes of chemical recycling →
  - depolymerization of PLA,
  - degradation of PLA.

8.4.2 *Bio-based Polyamides*

Bio-based polyamides are fully, or partially derived from renewable biomass and natural oils, such as castor bean oil (Fig. 8.14) (for example: the fuel lines of 95% of all North America Ford vehicles are made from castor oil-based polyamide 11) [23].

Examples of the application of bio-based polyamides in the automotive industry are shown in Table 8.6.

8.4.3 *Bio-based Polytrimethylene Terephthalate*

Bio-based polytrimethylene terephthalate (PTT) is a type of polymer derived from renewable biological sources (like sugar beets, corn, or other biomass feedstocks). It is a bio-polyester that exhibits remarkable properties such as a very good surface finishing, better mechanical properties at elevated temperatures, low warpage and good dimensional stability, making it a suitable alternative to conventional petrochemical-based polymers (bio-based PTT is biodegradable under certain conditions).

The automotive sector utilizes bio-based PTT for snap flit parts, knobs, and keyboards or instrument-panel air conditioning system outlet [31, 32].

Table 8.6 Examples of bio-based polyamides for the automotive industry [28–30]

Commercial name of polyamide	Characteristics as stated by the manufacturer
Nymax™ BIO Bio-based Polyamide (manufactured by: AvientCorp)	This bio-based nylon consists between 16 and 47% natural filler (corn, straw and wheat), and reduce the CO_2 value significantly at the beginning of the product lifecycle. They deliver comparable performance to prime PA66 glass fiber-filled materials, and these solutions have excellent surface appearance and color ability, and lower warpage. These formulations can be injection molded and extruded, making them suitable for automotive, consumer, industry, and construction applications. In addition, these biopolymers can be customized with laser welding and flame-retardant properties to meet additional performance demands Nymax BIO low water absorption PA formulations provide excellent dimensional stability and property retention after conditioning, solving the problem of water uptake for finished parts
Rilsan® polyamide 11 (manufactured by: Arkema)	The unique "11-carbon chain" yields a specific crystallinity profile that is very high in hydrogen bond density. This contributes to higher melting point, lower fuel/gas permeability, and improved impact properties compared to polyamide12. Polyamide 11 typically is more flexible, has better dimensional stability and better impact properties compared to typical short chain polyamides. It is easy to process, using most processing technologies (extrusion, extrusion blow-molding, injection molding). The polyamide11 matrix accommodates countless additives and filling agents. Polyamide 11 has a wide range of working temperatures (−40 °C to +130 °C). Rilsan® PA11 is used in various sectors worldwide, such as automotive, textile, oil and gas, and electronics

8.5 Recycled Plastics Used in Automotive Industry

Application recycled plastics in the automotive sector is a complex problem, consisting of several risk factors—ranging from questions related to quality of recycled materials, the price and volume of recycled materials (approx. 25% of plastic materials used in the automotive industry are recycled). Recently, however, companies have applied recycled plastic material to parts in interiors/exteriors [33, 34]:

- Škoda/Jaguar—application of PET bottles—used in production of rug fibers, floors and luggage components (Škoda Scala contains 14 kg of parts made from recycled platics) (Fig. 8.15),
- Hyundai/Jaguar—application of nylon waste from ocean (material is called Econyl®),

Fig. 8.15 Application of recycled plastics in Škoda Scala model (orange in pictures)

- Volvo—applied recycled and plastics from ship ropes and fishing nets (for production tunnel bracket),
- Ford—applied recycled plastics to create upholstery for passenger seat cushions in numerous models (for example, the seat fabric for each Focus is made with approx. 22 plastic water bottles),
- Continental (tire manufacturer): applied recycled polyester from plastic bottles (the PET bottles) into its tires productions,
- Reanult—efforts towards a so-called closed-loop economy by recovering polypropylene from the bumpers and wheel arch coverings so that it can be used to produce new plastic car parts.

Recycling regulations in the automotive industry: the research published by Field Action Science Reports, shows that in 2015 plastic derivatives represented 17% of the weight of cars (12% for polymers and 5% for elastomers) and that the forecast for 2030 indicates that this proportion will increase to almost 20% (15.6% polymers). The average weight of cars will be reduced from 1250 to 1123 kg, and they will contain an additional 25 kilos of plastic. The challenge is to significantly increase the amount of recycled materials [34].

When plastics reach their end-of life, these materials are treated to limit negative environmental impacts and to utilize the remaining value (Fig. 8.16). The treatment of plastics including plastics from end-of-live vehicles (ELVs) is divided into 4 options: reuse, material recovery, energy recovery, and landfilling (this may be the last option with relatively less environmental pollution).

Plastics recovery technology [35]:

- material recovery:
  - mechanical recycling (without significantly changing the chemical structure of the material = the polymer chains are not chemically disrupted in the process);

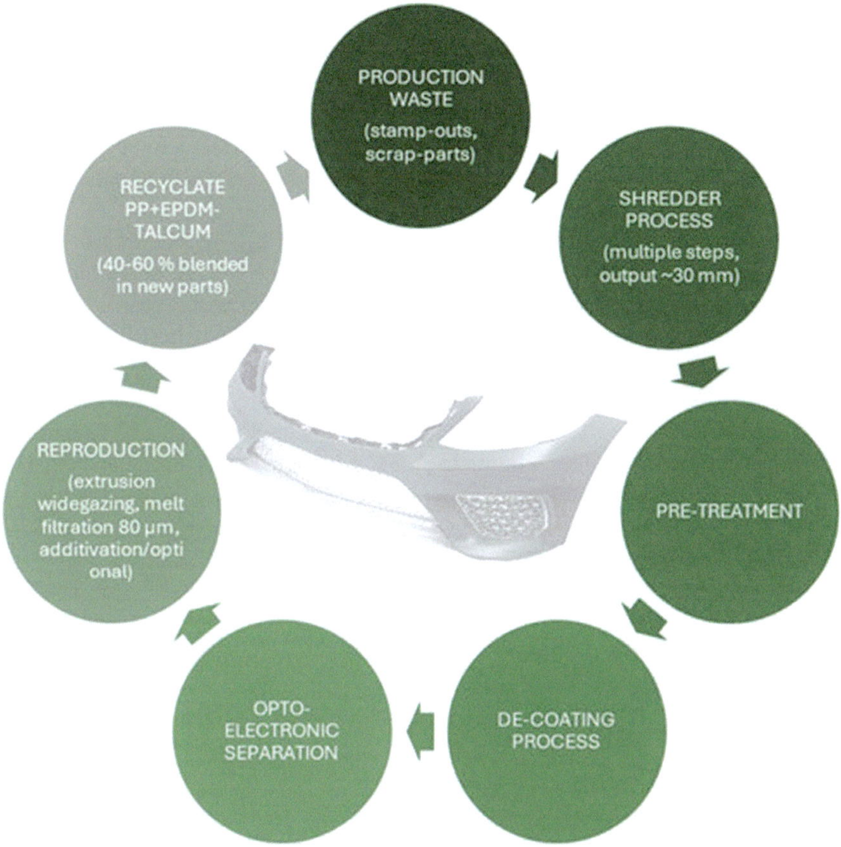

Fig. 8.16 Closed-loop-recycling (value chain: bumper → to bumper)

- chemical recycling (change the structure of polymeric waste and transforms it into chemical building blocks, including monomers, which can then be used as raw materials in chemical process. Chemical recycling includes processes such as gasification, pyrolysis, hydrocracking, and depolymerization);
- biological recycling (uses microbes to convert plastic waste into new products).

- *energy recovery* is the incineration of plastic waste, while simultaneously using the energy for generating electricity or steam or for providing process heat.

8.6 Summary

In recent years, the automotive industry has been focusing on reducing vehicle weight and improving fuel efficiency. One approach is to replace metallic materials and glass with plastics due to their advantageous characteristic properties. Plastics are among the most cost-effective materials, their production is relatively straightforward, and plastic components (such as bumpers, panels, dashboards, etc.) offer higher levels of structural flexibility. These materials have enabled significant vehicle design, manufacturing, and performance advancements, from early applications like celluloid to modern engineering plastics. The ability to tailor plastics to specific needs through additives and modifications has further expanded their utility, making them indispensable for various interior and exterior components.

The wide variety of plastics, polypropylene (PP), polyurethane (PUR), polyvinylchloride (PVC), acrylonitrile–butadiene–styrene (ABS), polyamide (PA), polyethylene (PE), polystyrene (PS), polyoxymethylene (POM), polycarbonate (PC), polymethyl methacrylate (PMMA), and polybutylene terephthalate (PBT), provides unique benefits tailored to specific automotive uses.

However, concerns arise regarding waste generated after the vehicles reach the end of their life cycle and its impact on the environment. This issue can be partially addressed by using recyclates, aiming to create a so-called closed-loop economy, or by employing biodegradable/degradable plastics (efforts known as the Bioconcept-Car).

References

1. Hutyrová Z, Litecká J, Harničárová M (2014) Technológia zmeny tvaru. FVT, Prešov
2. Plastics – the facts 2014. https://www.plastics.gl/market/plastics-the-facts-2014/. Accessed 25 Apr 2024
3. Mitaľová Z, Mitaľ D, Simkulet V (2023) Application of composite materials natural fibers in automotive industry–short review. In: Balog M, Iakovets A, Hrehova S (eds) EAI international conference on automation and control in theory and practice, Slovakia, February 2023. EAI/Springer Innovations in Communication and Computing. Springer, p 193. https://doi.org/10.1007/978-3-031-31967-9_15
4. Gu F, Hall P, Miles NJ (2016) Development of composites based on recycled polypropylene for injection moulding automobile parts using hierarchical clustering analysis and principal component estimate. J Clean Prod 137:632–643. https://doi.org/10.1016/j.jclepro.2016.07.028
5. The history, importance, and use of plastics in automobiles. https://innovativeautomation.com/the-history-importance-and-use-of-plastics-in-automobiles/. Accessed 20 Jan 2024
6. Szeteiova K (2010) Automotive materials plastics in automotive markets today. https://www.mtf.stuba.sk/buxus/docs/internetovy_casopis/2010/3/szeteiova.pdf. Accessed 30 March 2024
7. Automotive. The world moves with plastics. https://plasticseurope.org/wp-content/uploads/2021/10/20181019-Automotive-Booklet.pdf. Accessed 20 Feb 2024
8. Plastics in automotive construction - technology development and prospects (2021). https://knaufautomotive.com/plastics-in-automotive-construction/. Accessed 20 Feb 2024
9. Soybean car. https://www.thehenryford.org/collections-and-research/digital-resources/popular-topics/soy-bean-car/. Accessed 20 Feb 2024
10. Chen K-I et al (2012) Soyfoods and soybean products: from traditional use to modern applications. Appl Microbiol Biotechnol 96:9–22. https://doi.org/10.1007/s00253-012-4330-7
11. Maxwell J (1994) Plastics in the automotive industry. Elsevier, Cambridge
12. Types of plastics used in the automotive industry. https://www.rapiddirect.com/blog/types-of-automotive-plastics/. Accessed 20 May 2024
13. Mitaľová Z et al (2023) Matrices in wood plastic composites: a concise review. TEM J 12:1973. https://doi.org/10.18421/TEM124-07

14. Elvars B (ed) (2016) Ullmann's polymers and plastics. Wiley-VCH, Weinheim
15. Moritomi S, Watanabe T, Kanzaki S (2010) Polypropylene compounds for automotive applications. Sumitomo Kagaku 1:1–16
16. Adams ME, Buckley DJ, Colborn RE (1993) Acrylonitrile-butadiene-styrene polymers. iSmithers Rapra Publishing, Shawbury
17. Merillas B, Villafañe F, Rodríguez-Pérez MÁ (2022) A new methodology based on cell-wall hole analysis for the structure-acoustic absorption correlation on polyurethane foams. Polymers 14:1807. https://doi.org/10.3390/polym14091807
18. Polyamide 6 (Nylon 6). https://www.entecpolymers.com/products/resin-types/polyamide-6-nylon-6. Accessed 20 Apr 2024
19. What's the difference between LDPE and HDPE? https://www.essentracomponents.com/en-us/news/manufacturing/injection-molding/what-s-the-difference-between-ldpe-and-hdpe. Accessed 20 Apr 2024
20. Plastic materials index. https://precisionpunch.com/plastic-materials-index/. Accessed 20 Apr 2024
21. Bio-based plastics in the automotive market – clear benefits and strong performance. https://docs.european-bioplastics.org/2016/publications/fs/EuBP_fs_automotive.pdf. Accessed 20 Apr 2024
22. Altalhi T (ed) (2022) Handbook of bioplastics and biocomposites engineering applications. Wiley, New Jersey
23. What is PLA plastic? https://www.biopak.com/au/resources/what-is-pla. Accessed 05 May 2024
24. Tseng SCW (2012) Using bio-based materials in the automotive industry. Dissertation, University of Windsor
25. The future of car recycling is now. https://arn.nl/en/bio-based-plastic-is-emerging-for-the-automotive-industry/. Accessed 10 May 2024
26. Hill K, Swiecki B, Cregger J (2012) The bio-based materials automotive value chain. https://www.cargroup.org/wp-content/uploads/2017/02/The-Bio_Based-Materials-Automotive-Value-Chain.pdf. Accessed 15 Feb 2024
27. Biopolymers used in automobile industry. https://actgroup.ir/biopolymers-used-in-automobile-industry/. Accessed 15 May 2024
28. Rilsan® polyamide 11 (PA11). https://hpp.arkema.com/en/product-families/rilsan-polyamide-11-resins/. Accessed 15 May 2024
29. Biobased polyamides: bridging sustainability and performance. https://www.bio-sourced.com/biobased-polyamides/. Accessed 18 May 2024
30. Bio-based polyamide solutions. https://www.avient.com/products/engineered-polymer-formulations/sustainable-formulations/nymax-bio-bio-based-polyamide-solutions. Accessed 19 May 2024
31. Thielen M (2013) PTT. https://issuu.com/bioplastics/docs/bioplasticsmagazine_1301. Accessed 20 May 2023
32. Fernandes E (2024) Bio based polythrimethylene terephtalate market size, share analysis: projections of share, trends, and growth for 2024–2031. https://www.linkedin.com/pulse/bio-based-polytrimethylene-terephthalate-market-k1kpc/. Accessed 30 March 2024
33. Mitaľová Z, Dupláková D, Mitaľ D (2022) Application of recycled plastics in automotive industry: a short review. SAR J 5:200–205. https://doi.org/10.18421/SAR54-05
34. Plastics recycling in the strategies of well-known automotive brands. https://knaufautomotive.com/recycled-plastics-in-the-automotive-industry/. Accessed 20 May 2024
35. Schonmayr D (2017) Automotive recycling, plastics and sustainability. Springer, Graz

Chapter 9
Biocomposites

Abstract The chapter explores bio-composites, materials composed of natural-based reinforcements and resins, with at least one renewable component. It classifies bio-composites into three types: those with synthetic matrices and natural reinforcement, natural matrices and synthetic reinforcement, and entirely natural "green bio-composites." The chapter emphasizes the importance of polymer matrix selection based on thermal properties, as natural fibers degrade above 200 °C. It details the structure, composition, and treatment methods of natural fibers to improve adhesion and reduce moisture sensitivity. A significant focus is placed on applying natural fiber-reinforced polymers (NFRPs) in the automotive industry, noting their benefits in reducing weight and emissions and enhancing sustainability. The chapter also describes processing techniques for thermoplastic and thermoset matrices, highlighting the structure, properties, and production of wood-plastic composites (WPCs).

9.1 Introduction

Bio-composite materials are composite materials made from natural-based reinforcement and resins. To be considered a bio-composite, one of the main elements of the composite material must be a renewable resource.

Bio-composites are divided into three groups (where at least one component is defined as «bio»). In some cases, engineering plastics can also be applied [1]:

1. bio-composites with synthetic matrix and natural reinforcement,
2. bio-composites with natural matrix and synthetic reinforcement,
3. bio-composites with natural matrix and reinforcement (this category is called «Green bio-composites»).

In general, it is possible to find the following abbreviations in the literature: NFRC—Natural Fiber Reinforced Composite, NFRP—Natural Fiber Reinforced Polymer, resp. WPC—Wood Plastic Composite (with wooden reinforcement).

It is very important to choose correct polymer matrices for production of the bio-composites. The choice of the matrix depends on several factors; however, the

K. Berladir et al., *Materials for the Automotive Industry*, Springer Tracts in Mechanical Engineering, https://doi.org/10.1007/978-3-032-05866-9_9

primary selection criterion is the processing temperature—maximum 200 °C. At higher temperature, cellulose fibers are decomposed. The matrices in a NFRP can be divided into thermoplastic matrices or thermoset matrices. Thermosets are nonrecyclable polymers possessing stronger bonds and stiffness than thermoplastics. The most common thermosets used in NFRP are epoxy, polyester, and phenolic/common thermoplastic matrices used are: PE (polyethylene, or his form), PP (polypropylene), PS (polystyrene), and PVC (polyvinylchloride). In some cases, engineering plastics are also applicable (high performance segment of synthetic plastics with excellent thermal and mechanical properties in comparison to commonly used commodity plastics—e. g. ABS, PC, PA, etc.), or plastic recyclates.

In the following section of this chapter, natural reinforcements/fibers will be specified in more detail, as types of matrices are covered in the previous section of the publication titled «Plastics». A separate subchapter will focus on the application of bio-composites with natural reinforcement (referred to as types 1 and 3) in the automotive industry.

9.2 Natural Fibers

The term *natural fibers* (NFs) refers to fibers of vegetable or animal origin. Only cellulose fibers of plants obtained from various parts of these plants are used as reinforcement for the bio-composites in the automotive industry (Figs. 9.1, 9.2).

Natural reinforcements can be used in the form of long fibers/short fibers or crushed grains (example—some kind of wood flour in WPC products). A disadvantage of natural fibers is their poor durability—moisture absorption (moisture content can reach from 8 to 12.6%), subsequent swelling, susceptibility to ultraviolet light, their degrading temperature (approx. 180–200 °C), poor fire resistance and incompatibility with polymeric matrices (i.e., the resulting composite) [2].

However, the disadvantages can be suppressed by appropriate treatment methods, thus modifying the fibers' properties. The modification can be carried out by applying [4, 5]:

- *physical techniques* like plasma treatment and corona discharge can improve the functional properties of NFs. Low-temperature plasma treatments improve the surface of NFs by causing chemical implantation, etching, polymerization, free radical formation and crystallization. Plasma treatment is regarded as being an environmentally friendly and cost-effective technique, although some of the gases employed can be harmful to the environment;
- *chemical techniques* like:
  - alkali treatment is a common method to clean and modify the surface of natural fibers to promote enhanced fiber-polymer adhesion. Mercerization is a traditional alkali treatment based on sodium hydroxide (NaOH, caustic soda). A number of studies have found that alkali treatments can improve the properties of natural fibers and the interfacial adhesion to polymers;

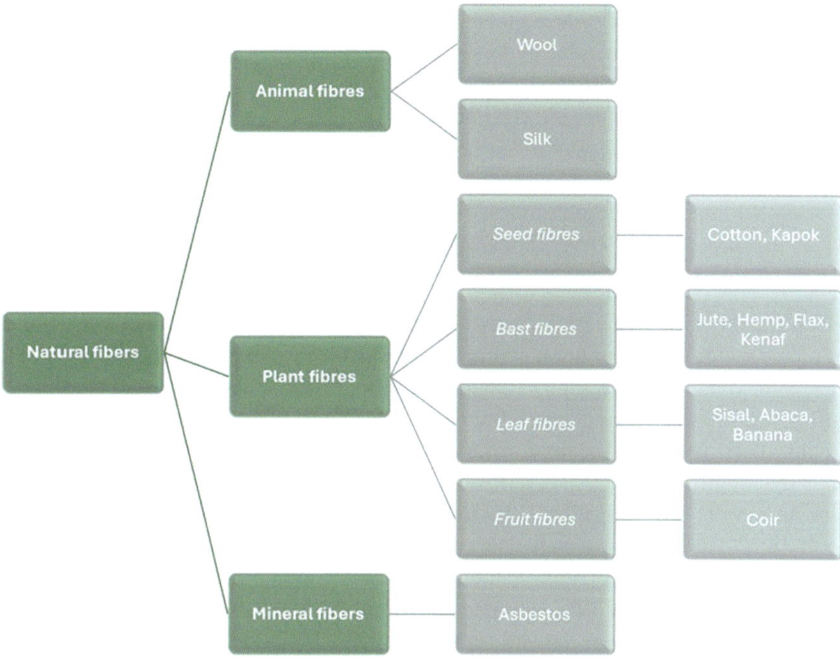

Fig. 9.1 Classification of natural fibers (+ examples of individual plant fibers)

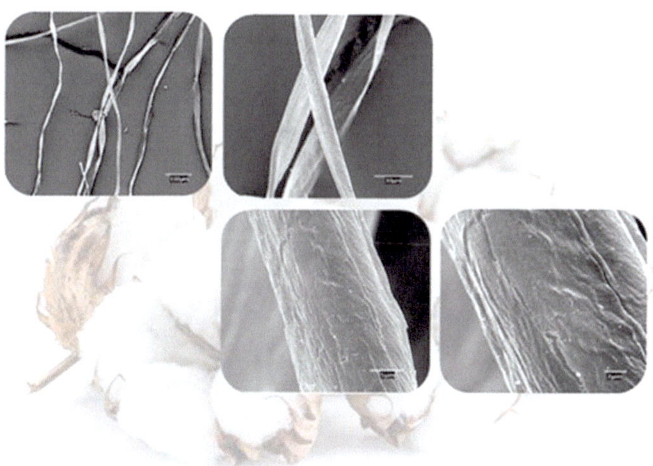

Fig. 9.2 Frutis of cotton plants and cotton fibers with different magnifications [3]

- acetylation is a procedure that entails the use of a suitable catalyst to eliminate atoms of hydrogen in the hydrophilic hydroxyl branches of cellulose molecules by grafting and placement of acetyl groups onto the cellulose structure at these initially hydrogenated sites. NFs can be acetylated with/without an acid catalyst to graft acetyl groups onto the cellulose structure. Acetylation helps to reduce the susceptibility of NF's to moisture absorption;
- by adding appropriate binders—for example isocyanates (strength and stiffness of wood-plastic composite can be increased by treating the fibers with polymethylene-polypheyl-isocyanate), or silane.

Natural fibers are materials derived from renewable resources of plants (lignocellulosic), animal and mineral origin. Whereas the animal fibers' main component is protein, plant fibers comprise components as follows: cellulose, hemicellulose, lignin, pectin, waxes, and other low-molecule substances (cellulose is the major structural element found in the form of lean rod-like crystalline microfibrils, aligned along the length of fiber). All cellulosic fibers are made up of filamentous units called microfibrils. These microfibrils assemble to form a fibrillar bundle which group together to form a single strand. The microfibrils in the densest S2 layer (secondary wall) are structured in a helical pattern and the angle at which they are positioned off the filament or fiber axis is called the → *microfibrillar angle* (the mechanical characteristics of the fiber depend on it) (Fig. 9.3). Smaller angles indicate greater stiffness/strength, larger angle signify greater ductility [6].

The mechanical properties and physical properties of natural fibers vary considerably depending on the chemical structural composition, fiber type, their aspect ratio, orientation, dispersion (will govern the material behavior during various stresses) and growth conditions [7].

Table 9.1 compares properties of some natural fibers and glass fibers. NFs have approx. ½ the density of glass fibers and have similar tensile modulus, but they have lower tensile strength a significantly higher moisture absorption. Natural fibers are hydrophilic in nature state which can lead to non-compatibility when combining with polymer matrix material (hydrophobic) [4].

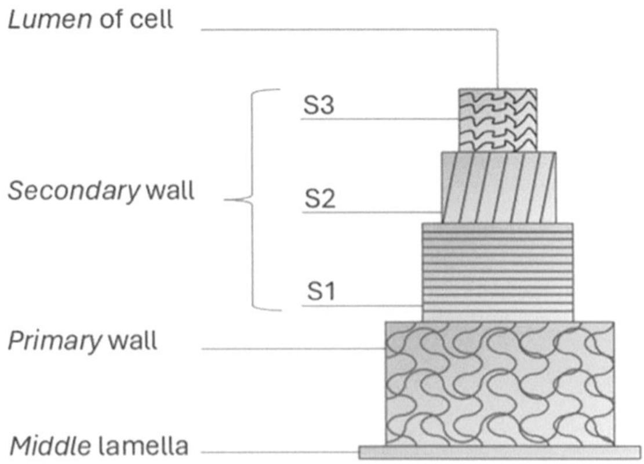

Fig. 9.3 Structure of natural fiber (microfibrillar angle—layer S2)

Table 9.1 Comparison of properties of selected types of natural fibers (compared to conventional reinforcing fibers) [4]

Type of fiber	Density (g cm^{-3})	Tensile strength (MPa)	Elongation to failure (%)	Tensile modulus (GPa)
Flax	1.4	500–900	1.2–1.6	60–80
Flax (description): belongs to the category of bast fibers. It is grown in temperature regions. Flax fibers possess characteristics such as a low density, high strength and stiffness, renewability				
Hemp	1.48	300–800	1.6	30–70
Hemp (description): belongs to the category of bast fiber (extracted from Cannabis plant family). Hemp fibers possess excellent mechanical strength, good insulation properties, resistance to salt water				
Jute	1.46	200–800	1.8	20–55
Jute (description): extracted from the ribbon of the stem of plants of genus Corchorus (bast fiber). Jute is the one of cheapest natural fibers. Jute fibers possess good thermal/insulation properties, high aspect ratio (l/d)				
Cotton	1–51	300–600	3–10	6–12
Cotton (description): belongs to the category of seed fibers. In nature, cotton is the purest form of cellulose containing 90%. It possesses excellent absorbency (it can absorb moisture up to 20% of its dry weight)				
E-glass	2.55	2400	3	73

9.3 Application of Bio-composites with Natural Reinforcement in the Automotive Industry

Recently the use of natural fibers in automotive industry has expended. It is reported that the increase of natural fibers in automotive components is expected (by the 54% per year). A number of automotive components (previously made with glass fiber composites) are now being manufactured using environmentally friendly composites (mechanical properties of natural fibers are much lower than those of glass fibers but their specific properties—especially—stiffness, are comparable to the glass fibers). The application of natural fibers to vehicle components can certainly reduce fuel intake and CO_2 emissions (a reduction of 10% in automobile weight improves fuel efficiency by approx. 7%). The increasing awareness of ecology and environmental sustainability, the more natural fibers are used because of their sustainable benefits, such as being environmentally friendly, more economical, and biodegradable. The major challenges of NFRP composites in automobile sector are low fire resistance, lack of adhesion between fiber versus matrix, high water absorption, and maximum processing temperature above 200 °C (for short period—it is possible to use temperature higher than T_{max}). Poor compatibility and high-water absorption of NFRP composites can be tackled by treatment method (as mentioned in the previous chapter).

Some facts from history of bio-composite applications: in 1941, Henry Ford developed the first prototype of a composite car referred to as a Soybean Car = … «*soybean*

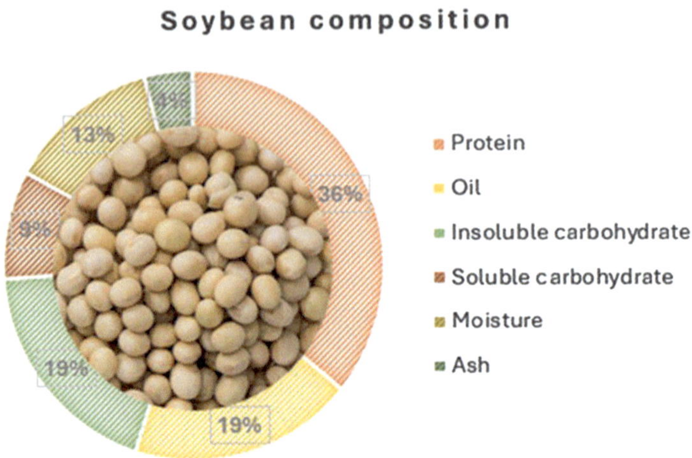

Fig. 9.4 Soybean composition

fiber in a phenolic resin with formaldehyde used in the impregnation» Davis, 1941 (cutting weight to 907 → 454 kg lighter than a steel car) (soybean composition: Fig. 9.4).

However, a breakthrough in the field of the NFRP application occurred later—in 1957 Trabant 601 was made in East Germany with a body made of duroplast (hardened plastic made of waste fibers of cotton + phenolic resins). Followed by Fiat punto (1982) with wood flour-filled door panels. Since the 1990s, there has been intense research in the field of replacement of original parts (whether metal or composite containing glass fibers) by composites reinforced with natural fibers (Table 9.2) [2].

Bcomp Ltd. (Fribourg, Switzerland) was nominated as one of the six candidates for the *«Biocomposites of the Year 2019»* award in 2019. It is dedicated to developing reinforcement for the NFRP composite materials. Their reinforcement was applied by automotive companies in the production of the parts as outlined below [2, 9]:

- VOLVO cars (XC60 passenger car parts)/Porsche: application of powerRibs™ fibers (source: hemp);

Table 9.2 LCA (life cycle analysis) of composite materials with natural fibers—in production [8]

Product (type)	Composite (type)	Traditional material	Weight reduction (%)	Change in greenhouse gas emissions (%)
Car side door	Hemp + EP	ABS	−27	−15
Under-floor pan	Flax + PP	GP + PP	0	Not specified
Rotor blade	Flax + PP	CF + EP	0	−45
Car interior	Bagasse + PP	Talc + PP	−20	−21

Table 9.3 Technical specification of product [10]

Technical specification	Composite*
Young's modulus in the way ribs direction	3.45 GPa
Young's modulus in the flat ribs direction	2.65 GPa
Shear modulus	280 MPa
Poisson's ratio	0.04
Density	363.6 kg m^{-3}
Thickness	1.21 mm

* These are properties of the equivalent composite layer which can be used to simulate the effect of a powerRibs layer infused with R&G resin L and GL2 hardener, 45% Vt of fibers

- McLaren Formula 1: application of AmpliTex™ to build of racing car seats (reduction of vibration, increased safety), for the first time at the 2021 British Grand Prix, LCA analysis reveals the seat has 85% lower CO_2 emissions.

  Product description: powerRibs™ (Table 9.3, Fig. 9.5) [10]:

- suitable for manufacturing reinforced composite product with high performance and low environmental impact (Fig. 9.6). Used to create a rib structure on one surface of the composite layer, significantly increasing the flexural stiffness and damping properties on thin composite shells with little additional weight;
- due to the discontinuous 3D structure, the mechanical properties of the powerRibs cannot be described in a generic way. The mechanical properties of product applied to a base layer can be represented by an equivalent layer represents the "smeared" effect of the powerRibs over the parts surface;
- note that equivalent modulus is not identical in the both in-plane directions. This is due the ribs being bonded flat to the base in one direction;

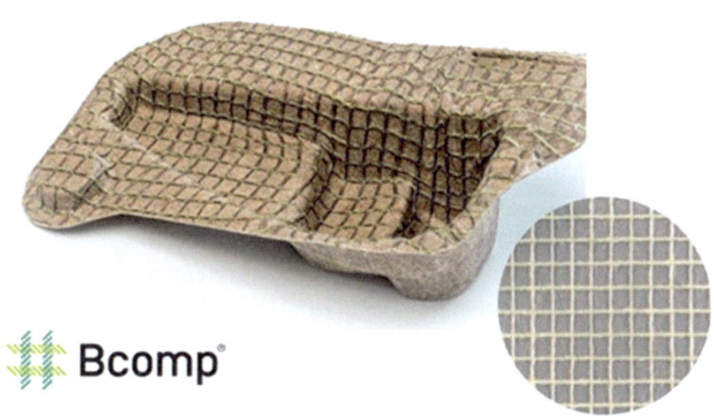

Fig. 9.5 Product PowerRibs™ (Bcomp®) [11]

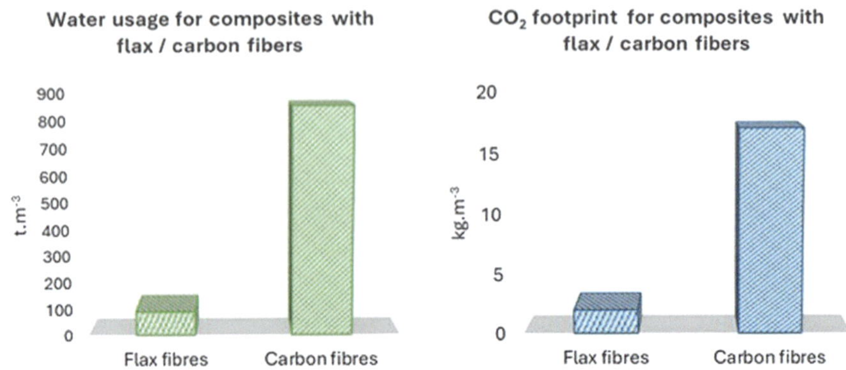

Fig. 9.6 Comparison of characteristics between flax and carbon fibers

- due to its grid structure with big tough yarns, powerRibs provides a very high damage tolerance to the part, also ensuring that it statys in one part even when crashed;
- compatible with epoxy, PP, polyester matrix;
- the key factor to optimize the effect of the powerRibs is to create a good rib structure;
- the ribs are normally used as last layer on the surface.

Faurecia (Nanterre, France) part of Forvia group—has been the market leader in sustainable vehicle interiors since the 1980s. Since 2003, it has focused on the application of the natural fiber polypropylene composite (in short NFPP), with approx. 45% of natural fibers → product: NAFILean (hemp fibers). Selected properties of product NEFILean-PF2 555 (20% hemp fibers reinforced PP compound for lightweight interior automotive parts) [12]:

- melt flow rate (190 °C, 5 kg/ISO 1133) = 11.6 g/10 min,
- density (ISO 1183) = 0.98 $g.cm^{-3}$,
- tensile modulus (ISO 527–2) = 2650 MPa,
- Izod impact notched (23 °C/ISO 180) = 7.5 $kJ.m^{-2}$,
- thermal properties—HDT A (ISO 75/A) = 72 °C.

Examples of companies producing composite materials with natural reinforcement (processed on the basis of the publication [2]): Johnson Control GmbH & Co. KG (non-visible door panel elements for BMW 3 series in 2021—made from plastic matrix + wood fiber), Bioplastics (plastic matrix + jute/hemp/sisal/flax/ American bamboo—for the production of components from door and trunk liners to under-floor pans to floor panels), Quadrant Natural Fiber Composites GmbH (50% PP + natural fibers → Quadrant's Nafcoform material, among the vehicles using this material are the AUDI A8, Mitsubishi Space Star Minivan, BMW 7-series, etc.).

Table 9.4 summarizes the typical application of biocomposites in the automotive industry.

Table 9.4 Application of composite materials with natural fillers (reinforcement) in automotive industry [13–19]

Company	Model	Material	Application
AUDI	Audi A2	Polyurethane filled with a blended sisal/flax fiber	Door trim panels
BMW	BMW 7-series	Thermosetting BASF acrylic copolymer and prepreg natural fiber mats	Door panels
Cadillac motor car devision	Cadillac DeVille	Wood fiber composites (specified further in the text below the table)	Seatbacks
FORD	Not specified	Wheat straw—as a filler for PP	Storage bin (injection-molded)
	Mercury Mariner/ Escape	Soy-based polyurethane	Seats, headliners, headrests
	Mustang GT RTD (BioConcept-Car)	Epoxy resin with flax fabric	Tailgate, rear spoiler, roof spoiler, closed underbody, diffuser, front splitter, bonnet
	Mondeo	Kenaf fiber/PP resins	Door panel
General motors	Chevrolet Impala 2002	Flax reinforced with PP	Trim panels, back shelf components
Honda	Pilot SUV	Wood fibers (matrix not specified)	Floor components
Lotus	Eco Eclise	Hemp-reinforced polyester	Body panels
Mercedes-Benz	Travego	Polyester/flax composites	Engine cages (to sound isolation)
	E-class	Jute fiber/epoxy matrix	Door panels
	S-class	Flax + hemp + sisal fibers/PU	Inner door panel
	Mercedes A-Class (W169)	Abaca fiber—as composite reinforcement	Spare wheel cover
Opel	Vectra	Blend of kenaf/flax (matrix not specified)	Package tray (parcel shelves), door panels
Renault	Renault 5 E-Tech	Hemp fibers/recycled PP and PET	Interior applications
The Toyota motor corp	Toyota Kijang Minibus	Kenaf fiber/phenolic resin	Flat door trim
Volkswagen	Not specified	Hemp/flax/kenaf—as composite reinforcement	Door trim
Volvo	C70/V70	Soya-based foam	Seat

The processing techniques used to fabricate NFRP composites on the type of polymer matrix used. For thermosetting polymers are suitable [19]:

- hand lay-up,
- RTM (Resin Transfer Molding),
- compression molding,
- filament winding,
- pultrusion,
- and vacuum-assisted resin transfer molding.

On the other hand—for thermoplastic polymers are suitable processes such as injection molding, extrusion, thermoforming, and compression molding. In addition, processing of thermoforming composite depends also on fiber length (for example: injection molding is preferable for short fiber thermoplastic composites, extrusion/compression molding for long fibers, and thermoforming/pultrusion/thermostamping/filament winding for continuous fibers) [19].

Wood-plastic composites (WPCs) are created by combining polymer matrix, wood flour and suitable additives (in certain reaction) (Fig. 9.7). The main function of modifiers is to improve mechanical properties, provide a chemical stability and easier mixing of components. The presence of wood in plastic matrix increases strength and stiffness. WPCs are environmentally friendly because are produced from agricultural waste and recycled plastic (for example: PVC bottles and carpets). As a filler typically used in WPC products, a wood flour obtained from maple, oak, pine, etc. (according to the location of manufacturers). Wood flour presents crushed wood grains, and their appearance is similar to standard flour. There are two steps of wood production: size reduction (using hammer mill or chipper), and size classification (screening). The particles are sorted by size, they run through the sieves with different mesh and consequently are classified according to the U.S standard (for example: a particle with diameter 850 μm with identification of 20 mesh according to the standard). During the melting process it is necessary to ensure that the temperature does not exceed 200 °C, because there would be a wood decomposition. The crushed mixture consists of fibers with aspect ratio: length divided by diameter (l/d). This value ranges from 1 to 5 (a lower aspect ratio helps to provide a simple transport into the extruder but decreases mechanical properties) [20].

Depending on the processing temperature, WPCs are suitable as matrices: various forms of polyethylene (PE, LDPE, HDPE, etc.), polypropylene (PP), polyvinyl chloride (PVC), polystyrene (PS); thermosets: phenolic, epoxy and polyester resins. Studies are currently underway to verify the use of polymethyl methacrylate (PMMA) and Nylon 6 and biodegradable polymers.

Additives are important ingredients of WPCs that can be used for various purposes. The types of additives used in the WPC industry include coupling agents (to improve compatibility of between fiber and matrix), lubricants (non-soluble or semi-soluble lubricants), flame retardants (to reduce a risk of ignition of the product), biocides (to protect the organic components), pigments (to create a character of surface), fillers (to reduce the prices of material).

Fig. 9.7 Wood shaving, wood flour and pellets (mixture of matrix and fillers)

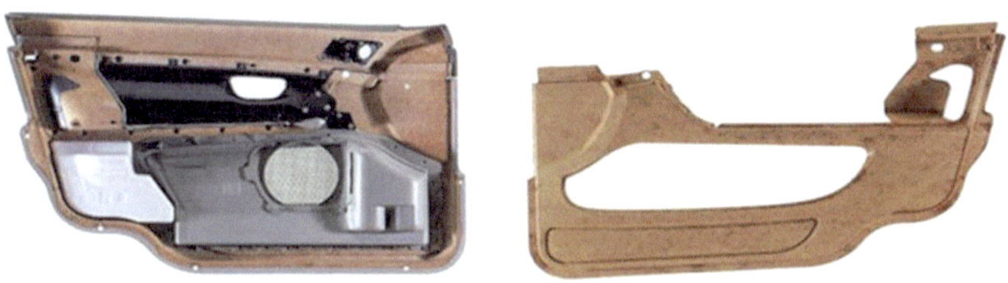

Fig. 9.8 Door panel interior made from wood fiber reinforced composites [21]

The primary manufacturing processes for WPCs are extrusion, injection molding, and compression molding or thermoforming (pressing). Newer manufacturing processes for WPCs include additive manufacturing via fused layer modeling and laser sintering. Properties of WPC materials depend on several aspects and final application (for example, Fig. 9.8). The material can be customized according to customer requirements [20].

9.4 Summary

EU regulations on emission limits are leading automobile manufacturers towards vehicle construction lightweighting. One option is the application of composite materials with natural fibers. These composite materials can be tailored to achieve good insulation, thermal, and acoustic properties, high strength, and stiffness, while maintaining lower density (and weight), which is critical in automobile production to maintain lower fuel consumption.

Bio-composite materials are composed of natural-based reinforcements and resins, with at least one component derived from renewable resources. These materials are categorized into three types: bio-composites with synthetic matrices and natural reinforcement, bio-composites with natural matrices and synthetic reinforcement, and green bio-composites, which use both natural matrices and reinforcements. The selection of polymer matrices is crucial in bio-composites, as they must withstand processing temperatures without degrading cellulose fibers. Thermoplastic and thermoset matrices are commonly used, each offering different benefits such as recyclability, stiffness, and durability. Natural fibers, typically derived from plant cellulose, are used in various forms but present challenges such as moisture absorption and poor fire resistance, which can be mitigated through treatments like plasma or chemical modifications.

Historically, early bio-composite applications in vehicles date back to Henry Ford's 1941 soybean car, but significant advancements occurred in the late 20th century with models like the Trabant and Fiat Punto incorporating natural fiber composites. Leading automotive brands today, such as Volvo, McLaren, and Mercedes-Benz, actively integrate these materials into their vehicles, leveraging their lightweight and sustainable properties.

Composite materials with natural fibers are also suitable for recycling. Studies indicate recyclability approximately 4–6 times until their thermomechanical properties change (the recycling process depends on the components). Given their wide range of advantages, they become an alternative to carbon fiber composite materials.

References

1. Altalhi T (ed) (2022) Handbook of bioplastics and biocomposites engineering applications. Wiley, New Jersey
2. Mitaľová Z, Mitaľ D, Simkulet V (2023) Application of composite materials natural fibers in automotive industry–short review. In: Balog M, Iakovets A, Hrehova S (eds) EAI international conference on automation and control in theory and practice, Slovakia, February 2023. EAI/Springer innovations in communication and computing. Springer, p 193. https://doi.org/10.1007/978-3-031-31967-9_15
3. Serra A et al (2021) Exploring the potential of cotton industry byproducts in the plastic composite sector: macro and micromechanics study of the flexural modulus. Materials 14:4787. https://doi.org/10.3390/ma14174787
4. Quarshie R, Carruthers J (2021) Technology overview biocomposites. https://compositesuk.co.uk/wp-content/uploads/2021/12/Biocomposites-Tech-overview-2014.pdf. Accessed Jan 28 2024
5. Oladele I et al (2020) Acetylation treatment for the batch processing of natural fibers: effects on constituents, tensile properties and surface morphology of selected plant stem fibers. Fibers 8:73. https://doi.org/10.3390/fib8120073
6. Uppal N et al (2007) Cellulosic fibres-based epoxy composites: from bioresources to a circular economy. Ind Crops Prod 182:114895. https://doi.org/10.1016/j.indcrop.2022.114895
7. Saira T, Munawar A, Shafiuallah K et al (2007) Natural fiber-reinforced polymer composites. https://www.researchgate.net/publication/228636811_Natural_fiber-reinforced_polymer_composites. Accessed 30 May 2024

8. Duflou JR et al (2012) Do fiber-reinforced polymer composites provide environmentally benign alternatives? A life-cycle-assessment-based study. MRS Bull 37:374–382. https://doi.org/10.1557/mrs.2012.33
9. The next step in McLAREN's use of sustainable composites in F1. https://www.mclaren.com/racing/sustainability/sustainable-natural-fibre-composites/. Accessed 15 May 2024
10. powerRibs. https://pdf.directindustry.com/pdf/bcomp/powerribs/201565-801479.html. Accessed 20 Apr 2024
11. Mathijsen D (2018) The renaissance of flax fibers. Reinf Plast 62:138–147. https://doi.org/10.1016/j.repl.2017.11.020
12. Biomaterials for real. https://www.apm-planet.com/downloads/. Accessed 15 Apr 2024
13. Koronis G, Silva A, Fontul M (2013) Green composites: a review of adequate materials for automotive applications. Compos B Eng 44:120–127. https://doi.org/10.1016/j.compositesb.2012.07.004
14. Boland CS et al (2016) Life cycle impacts of natural fiber composites for automotive applications: effects of renewable energy content and lightweighting. J Ind Ecol 20:179–189. https://doi.org/10.1111/jiec.12286
15. Sanyang ML et al (2016) Recent developments in sugar palm (Arenga pinnata) based biocomposites and their potential industrial applications: a review. Renew Sustain Energy Rev 54:533–549. https://doi.org/10.1016/j.rser.2015.10.037
16. Stewart R (2010) Automotive composites offer lighter solutions. Reinf Plast 54:22–28. https://doi.org/10.1016/S0034-3617(10)70061-8
17. Habermann Ch (2013) Mobility for the future. https://epaper.renewable-carbon-plastics.com/html5/MYcjTJt0lb/0zWSvN2qIEeDU/12. Accessed 20 May 2023
18. Ramli N et al (2018) Natural fiber for green technology in automotive industry: a brief review. In: IOP conference series: materials science and engineering, Beijing, 2017
19. Chauhan V, Kärki T, Varis J (2022) Review of natural fiber-reinforced engineering plastic composites, their applications in the transportation sector and processing techniques. J Thermoplast Compos 35:1169–1209. https://doi.org/10.1177/0892705719889095
20. Hutyrová Z (2012) Príspevok ku riešeniu problematiky trieskového obrábania kompozitných materiálov na báze WPC. Dissertation, Technická Univerita Košice
21. Patel RV, Yadav A, Winczek J (2023) Physical, mechanical, and thermal properties of natural fiber-reinforced epoxy composites for construction and automotive applications. Appl Sci 13:5126. https://doi.org/10.3390/app13085126

Chapter 10
Glass

Abstract The chapter explores the evolution, types, and integration of glass in the automotive industry, emphasizing its critical role in safety, comfort, and performance. Initially absent from early vehicles, glass became essential by 1904 and advanced to include laminated and tempered varieties. Laminated glass, primarily used in windshields, features dual panes bonded with polyvinyl butyral (PVB) for impact resistance and UV protection. Tempered glass, used in side and rear windows, is thermally strengthened to shatter into blunt fragments. Additional innovations include solar glass for heat reduction, heated glass for defrosting, and XIR glass for enhanced durability and operator comfort in heavy machinery. The chapter also details modern glazing systems, which streamline manufacturing via encapsulation, extrusion, bonding, and assembly processes. Materials are evaluated for their sealing, insulation, and durability properties. Overall, the chapter highlights the transformation of glass from a basic wind barrier to a multifunctional safety and design component in modern vehicles.

10.1 Introduction

Glass is one of the most important parts of any vehicle. The use of automotive glass in manufacturing car windshields, side windows, and rear windows has greatly increased safety. It also protects against harmful infrared radiation, prevents excessive car interior heating, and reduces noise levels.

Automotive glass is a specialized type used in vehicles, including windshields, side windows, and rear windows (Fig. 10.1).

Cars were originally designed without windshields. This is probably because the cars didn't go fast enough to justify the need for windshields. It wasn't until 1904 that windshields appeared on cars for the first time, 10 years after the first car was introduced [1]. The first windshields were nothing more than a piece of glass made from the same material that house windows are made of. They consisted of two cups, an upper and a lower part. They were designed so the top half could be folded down if needed [2].

K. Berladir et al., *Materials for the Automotive Industry*, Springer Tracts in Mechanical Engineering, https://doi.org/10.1007/978-3-032-05866-9_10

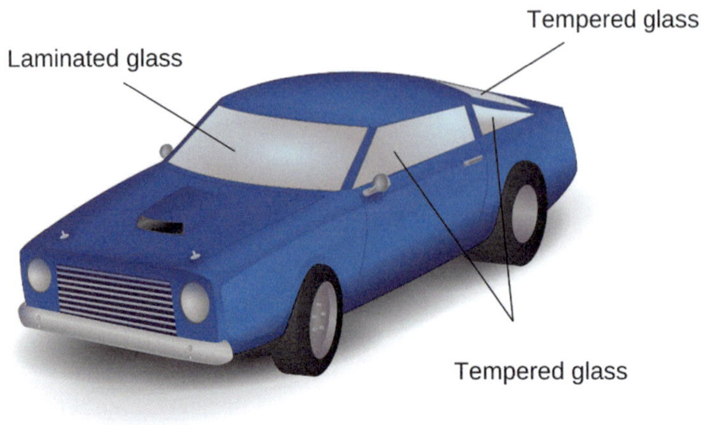

Fig. 10.1 Types of glass in car

Early windshields and other glass tended to shatter and were very dangerous. Fortunately, technology has improved; cars now use laminated and toughened glass. Laminated glass, originally used for the front and rear windows (tempered glass was used for the side windows), is now used throughout the vehicle. This practice started with luxury cars but continued with more mainstream cars.

Laminated glass is specially designed to bend under pressure and not break easily. It is also designed to reduce UV radiation, which is crucial when riding on a sunny day. Tempered glass is similar to laminated glass but slightly stronger and breaks differently [3]. It is designed to break into small pieces to reduce the risk of injury in the event of an accident. It was not until 1934 that laminated glass was first used in automobiles, and 15 years later, tempered glass also appeared for the first time. Technology has improved over the years and is now as good as ever.

Wraparound glass is no longer used in cars but was very popular until the 1950s. The intention was to reduce the blind spot of drivers and thereby increase safety. Wraparound windshields were one giant piece of glass encompassing the windshield and wrapped around the driver and passenger doors.

It wasn't until 1966 that the government finally started setting car regulations. The Highway Safety Act of 1966 was the first legislation to allow the government to set safety standards for automobiles. It was this federal law that made crash-resistant windshields mandatory [1].

Power windows were another revolutionary product that entered the automotive scene and took it by storm. Surprisingly, electric windows date back to the 1940s but were considered a luxury and were not fitted to all cars. However, one by one, car manufacturers began adding power windows to more vehicles. Finally, in the late 1970s, car power windows became the norm rather than the exception.

10.2 Types of Automotive Glass

10.2.1 Laminated Glass

Laminated glass is one of the most common types of glass used in vehicles. It comprises a glass panel made from two separate panes glued together using PVB (polyvinyl butyral) and bonded at high temperatures (Fig. 10.2).

The result is an incredibly strong type of glass that withstands impacts much more effectively than standard glass. In addition, when the car is involved in an accident, the laminated glass will not break. Although it may break on impact, the PVB layer keeps it intact so sharp glass shards won't fly off and cause serious injury [5].

Laminated glass can also help reduce collision injuries by acting as a barrier. If a passenger is thrown into the glass, it has a cushioning effect that can help reduce the severity of injuries sustained.

By law, all windshields are made of laminated glass. Although the side and rear windows may not be made of laminated glass, more cars are using this type of glass in these windows these days. In addition to the windshield or windshield, laminated glass is widely used in automobile sunroofs. A laminated sunroof offers greater safety, visibility and the ability to add tinting options for your vehicle. Some premium cars, such as the Mercedes Benz S350, use laminated glass, including the door windows and rear windshield.

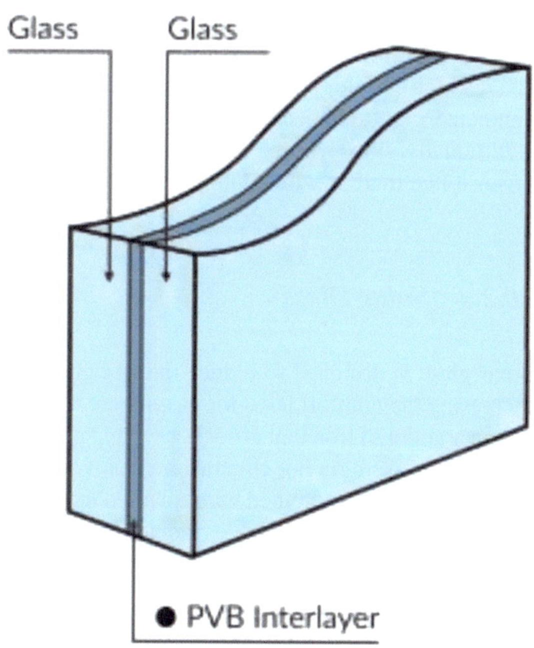

Fig. 10.2 Structure of laminated glass [4]

The main advantage of laminated glass in cars is that it can withstand serious accidents. It can be used on front and rear windows, providing excellent protection for drivers and passengers. Laminated glass can also be installed inside door windows for added security. The most common disadvantage of laminated glass is that it is prone to breakage in impact even with low force, and therefore, the cost of replacing a new windshield in car maintenance is very high.

10.2.2 Tempered Glass

Tempered glass is the other main type of automotive glass used today. It is four times stronger than standard glass. Tempered glass consists of a single layer of glass heated to more than 600 °C [3]. It then goes through a rapid cooling process called tempering, resulting in a much stronger glass.

In addition to strength, tempered glass has one more important advantage. Although it shatters on a strong impact, it shatters into many small, blunt-edged pieces rather than dangerous sharp glass shards, reducing the chance of serious injury from flying sharp glass shards. It is often used in rear and side windows and sunroofs, but some car models are now equipped with laminated glass [6].

The main advantage of tempered glass is its affordability. Compared to laminated glass, tempered glass is cheaper and easy to maintain. The low cost is why many commercial operators still use toughened glass in their buses, trucks, and RVs. In addition, tempered glass is stronger than laminated glass and is, therefore, less prone to shattering from rocks or debris. The main disadvantage is the absence of a PVB layer, which can stop the penetration of harmful objects into the car's interior. Unlike laminated glass, tempered glass shatters on impact and cannot protect occupants. Additionally, since tempered glass shatters completely on impact, there is no way to repair it like there is with a laminated windshield [7].

10.2.3 Solar Glass

Solar glass is designed to reduce the amount of heat entering the vehicle, thereby increasing the comfort level for passengers. Often used for windscreens, it contains a high amount of iron that absorbs heat and prevents it from entering the vehicle [8], particularly useful in hot climates such as Western Australia.

Standard glass combined with solar technology looks like regular glass but allows light to pass through and reduces radiation at the same time. The Low-E coating (Low-E) is invisible and is usually a permanent coating that does not require replacement.

10.2.4 Heated Glass

Heated glass is heated electronically and has long been used in vehicles. It usually consists of glass containing thin wires that can quickly reach high temperatures [9].

When the driver turns on the heater, the wires heat up immediately, melting the ice and helping to defrost the windows. It is most often used for rear windows but can also be found on front windows. It is mainly used for convenience. By helping to defrost the windows, it can also help improve visibility, making driving safer.

Another type of heated glass is where a clear, electrically conductive coating is applied to the glass, and then a current is passed through it to generate heat and heat the glass, with the same results. This is sometimes used for both front and side windows.

10.2.5 XIR Glass

XIR is an automotive glass designed to provide safety and heat resistance in automobiles. It is used primarily in heavy equipment, which provides operators with increased comfort thanks to its excellent heat resistance [10].

XIR glass is stronger and more durable than laminated glass and has many advantages, including increasing the safety of operators when using the equipment. Because it reduces solar heat gain and has noise-absorbing properties, it increases comfort and reduces fatigue.

Its durability means fewer glass changes are needed, and machine downtime is reduced, leading to increased productivity.

10.3 The Glazing System

Increasingly, vehicle manufacturers are sourcing glazing systems instead of solid glass. Much of this added value occurs after the glass is laminated or tempered by manufacturers. Prefabricated glazing systems help simplify the vehicle manufacturing process.

There are 4 different glazing processes. Modular systems such as encapsulation and extrusion are designed to make bonding the glass to the vehicle easier. The other two, gluing and assembly, add another step to the manufacturing process [11].

10.3.1 The Glazing Processes

- Encapsulation. It involves the injection of polymer cut-outs that are precisely shaped around the perimeter of the glass to match the vehicle's body. It also offers the option to include other styling elements such as fasteners, trim, braces, and even window curtains that are typical of some vehicles, with the same removal procedure. It has many advantages: it improves airflow, reduces water and dirt entry points, reduces internal noise, and improves the appearance of the final design.
- Extrusion. Plastic extrusion is an industrial process in which plastic is pressed and shaped using an extrusion machine. Using pressure and pushing, the desired shape is created through the mold. This is usually done using PVC or recyclable plastic. The equipment is heavy, bulky, and expensive. This is a more lengthy process because the material needs to be connected to the glass. In addition, there is a risk of producing material with an uneven texture or appearance.
- Bonding. The material (PVC or other) used to adhere to the glass uses adhesives and fasteners, adding another step to the manufacturing process. Additional steps in the production process increase the total cost of producing the product.
- Assembly. Fixing the glass in the car can also be achieved by installing clamps on the glass. The fastening of such clamps together with other fasteners is usually called an assembly. The plastic for fixing the glass is made earlier. Additional materials (clamps) are required. This must be done by the vehicle manufacturer (additional process).

10.3.2 The Glazing Materials

There are different materials used in the glazing process [2, 7, 11]:

- High-density polyurethane (PUR RIM**).** Polyurethane-encapsulated glass provides a better seal than glass. Unlike most solutions with two or three separate components, this glass encapsulation solution is a single unit including glass, gasket and spacer rod. In addition to being lightweight, it provides high sound insulation and an airtight seal. Glass encapsulation without secondary adhesives reduces component count and installation time. Polyurethane is a better thermal insulator and does not cause heating problems like PVC. In addition, its processing does not cause so many problems due to degassing. It is significantly less aggressive. Another note is that polyurethane is less susceptible to microbiological attack and weathering than rubber materials (TPE, EPDM). In addition, it has greater thermal and chemical resistance.
- Polyvinyl chloride (PVC). Thanks to its good price/quality ratio and great versatility, it is one of the most used plastics. It has good sealing properties, resists water and dust, and sound insulation. PVC encapsulation is often the most cost-effective solution for small glass components such as side or square windows

in automobiles. However, the PVC molding process presents several challenges. The outgassing of hydrochloric acid and hydrobromic acid causes discomfort to operators and affects molds and nearby equipment. In addition, it has a high risk of burns. PVC is not a good insulator and gets very hot during processing (it takes a long time for parts to cool after removal).

- Thermoplastic elastomer (TPE). Thermoplastic elastomer or thermoplastic rubber is a material that can be processed using thermoplastic technologies such as injection molding, 2 K molding, or extrusion. They are easily shaped and subject to further processing (painting, gluing). TPE has properties similar to rubber. Compared to thermoset rubbers (also RIM polyurethane), they have lower thermal and chemical resistance and worse yield (shrinkage under compression) after loading.
- Polyethylene propylene diene monomer rubber (EPDM). EPDM is an elastomer with very good properties against water penetration and atmospheric influences. It has very high elasticity and resistance, which makes it a suitable material for waterproofing all surfaces. EPDM has good elasticity, which allows it to resist tearing caused by other materials. However, it is easily damaged when cutting. It has less strength and impact resistance than PUR RIM. This material is more difficult to adhere to glass due to its waxy feel.

10.4 Summary

Glass has become an essential component of modern vehicles, contributing to safety, comfort, and energy efficiency. Initially, cars lacked windshields, but by 1904, simple glass panels were introduced, which later evolved into laminated and tempered glass for enhanced safety. Laminated glass, used mainly in windshields, consists of two glass layers bonded with polyvinyl butyral, preventing shattering and offering UV protection. Tempered glass, used for side and rear windows, is heat-treated to increase strength and break into blunt fragments upon impact, reducing injury risks. Other specialized types of automotive glass, such as solar glass for heat reduction, heated glass for defrosting, and XIR glass for enhanced durability, have further improved vehicle functionality and comfort.

The glazing system in modern vehicles has advanced significantly, incorporating processes such as encapsulation, extrusion, bonding, and assembly to streamline manufacturing and improve glass integration. Encapsulation injects polymer moldings around the glass for better sealing and aesthetics, while extrusion shapes plastic materials to fit the glass edges. Bonding and assembly techniques use adhesives and fasteners to secure glass panels, ensuring structural integrity. These methods enhance aerodynamics, reduce noise, and improve insulation. With growing safety and efficiency demands, manufacturers have shifted toward prefabricated glazing systems to optimize production and enhance vehicle performance.

Various materials are used in glazing systems, each with distinct properties affecting durability, insulation, and cost-effectiveness. High-density polyurethane offers excellent sealing, sound insulation, and lightweight advantages, making it a popular choice. Polyvinyl chloride is cost-effective but presents challenges such as heat retention and emissions during processing. Thermoplastic elastomer provides flexibility but has lower thermal and chemical resistance, while EPDM rubber excels in water resistance but lacks impact strength. Over the decades, automotive glass has transformed from a simple wind barrier to a critical safety feature, significantly improving passenger protection and driving experience.

References

1. The history of auto glass. https://glassdoctor.com/blog/the-history-of-auto-glass. Accessed 26 Feb 2025
2. The evolution of windshield technology: from safety glass to smart displays. https://pickuptrucktalk.com/2025/02/the-evolution-of-windshield-technology-from-safety-glass-to-smart-displays/. Accessed 26 Feb 2025
3. Gy R (2021) Tempered and laminated glazing for cars. In: Richet P, Conradt R, Takada A, Dyon J (eds) Encyclopedia of glass science, technology, history, and culture. Wiley, USA. https://doi.org/10.1002/9781118801017.ch9.2
4. Ahani A, Ahani E (2023) An overview for materials and design methods used for enhancement of laminated glass. Hybrid Adv 3:100063. https://doi.org/10.1016/j.hybadv.2023.100063
5. Nicolais L, Zang MY, Chen SH (2011) Laminated glass. In: Nicolais L (ed) Wiley encyclopedia of composites. Wiley, USA. https://doi.org/10.1002/9781118097298.weoc121
6. The different types of windscreens. https://www.thewindscreenco.co.uk/help-advice/different-types-windscreens/. Accessed 26 Feb 2025
7. The different types of glass used in automobile. https://windshieldstore.in/blog/the-different-types-of-glass-used-in-automobile/#:~:text=Tempered%20glass%20is%20widely%20used,to%20the%20laminated%20rear%20windshield. Accessed 26 Feb 2025
8. Solar control glass: the ultimate car windscreen solution (2016). https://www.aisglass.com/blog/solar-control-glass-the-ultimate-car-windscreen-solution/. Accessed 26 Feb 2025
9. What Cars Have Heated Windscreens? (2020) https://www.stablevehiclecontracts.co.uk/blog/what-cars-have-heated-windscreens/?srsltid=AfmBOopamU1HAzTgH_Pgp8ktJ1bCgavh8uX5-H5gtIuKZxqSSqfcHjzS Accessed 26 Feb 2025
10. Understanding the different types of auto glass. https://dcautoglass.com.au/understanding-the-different-types-of-auto-glass/. Accessed 26 Feb 2025
11. Materials and types of glazing systems in the automotive industry (2020). https://concentrol.com/de/glazing-systems-automotive-industry/. Accessed 26 Feb 2025

Chapter 11
Rubber and Rubber Materials

Abstract The chapter explores the critical role of rubber and rubber materials in the automotive industry. Beginning with their historic introduction through John Dunlop's pneumatic tire in 1888, the chapter details how rubber evolved from basic tire applications to essential components such as engine mounts, vibration isolators, seals, hoses, and gaskets. The transition from natural to synthetic rubbers improved durability, heat resistance, and chemical stability. Rubber's performance is further enhanced through vulcanization and additives such as carbon black and silica. Detailed classifications are provided based on composition, temperature tolerance, and function. The chapter highlights rubber's mechanical properties, including elasticity and vibration damping, and its susceptibility to aging under environmental stress. Modern developments include bio-rubber research and recycling, aligning with sustainability goals. The application-focused sections cover radial seals, brake system cuffs, hoses, liquid gaskets, and vibration isolation products, underscoring rubber's indispensable function in automotive engineering.

11.1 Introduction

Rubber and rubber products are used for various purposes in the automotive industry [1] (Fig. 11.1). In the automotive industry, the presence of rubber is observed in fluid transfer systems, body sealing systems, transmission systems, automotive anti-vibration systems, sealants, adhesives and coatings, O-rings, molded parts, flat seals, foam and convertible products, body parts, spare parts, etc. [2].

The automotive industry is directly related to the rubber industry. The growth of the industry and the importance of rubber go hand in hand. It is known that 75% of the rubber produced in the world is used to make tires [3]. The automotive industry dominates the rubber industry directly and indirectly. Direct interference refers to original equipment tires and many other elastomeric parts installed on the vehicle during production. In contrast, indirect interference refers to aftermarket parts such as tires, windshield wiper blades, and, to a lesser extent, other parts. Tires, engine

K. Berladir et al., *Materials for the Automotive Industry*, Springer Tracts in Mechanical Engineering, https://doi.org/10.1007/978-3-032-05866-9_11

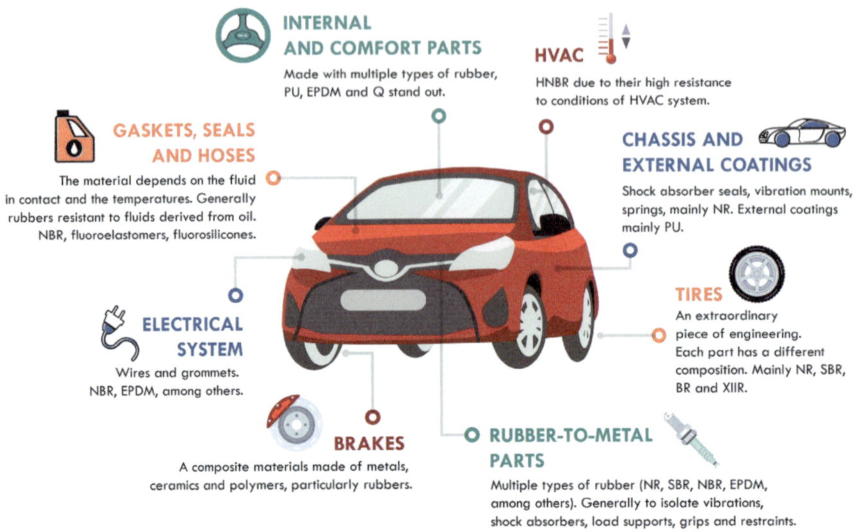

Fig. 11.1 Common rubber parts in a passenger car [1]

mounts, and windshield wiper blades are usually natural rubber. The raw materials used in certain transport industries are shown in Fig. 11.2 [4].

Many other parts in the automotive industry are made from synthetic rubber, although more natural rubber can be used (in combinations) in extruded sealants

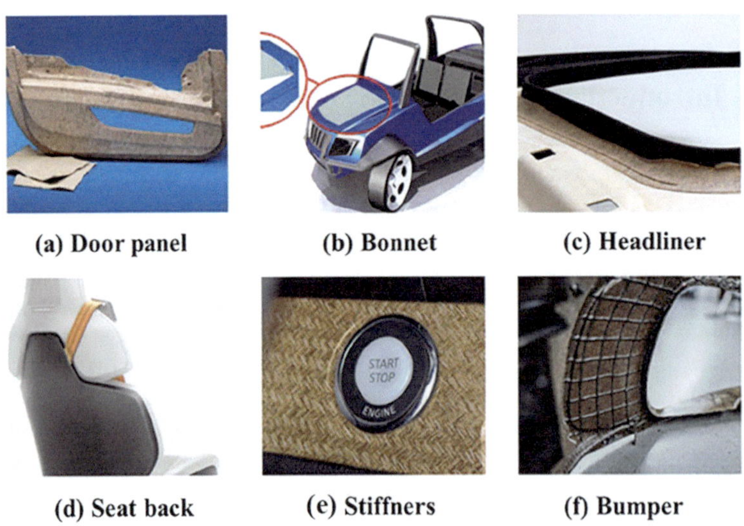

Fig. 11.2 Incorporation of natural fibers in automobile uses: **a** hemp, **b** jute, **c** kenaf, **d** flax, **e** jute, and **f** flax [4]

[5]. Some parts used in the engine compartment are made of high-quality synthetic rubbers, such as the air conditioning hose, engine seals, etc.

The use of rubber in cars began with tires. John Boyd Dunlop's invention of the rubber tire in 1888 marked a significant shift away from the metal wheels of early automobiles [6]. With an air-filled natural rubber construction, the tire provided shock absorption and traction, making the ride safer and more comfortable. Although a significant improvement over metal alternatives, natural rubber wore out relatively quickly and became hard and brittle at low temperatures. This initial Dunlop application influenced several advances in tire technology from other manufacturers. This included increased durability, performance in various road conditions, and wear resistance. These efforts led to the development of specialty tires designed for certain weather conditions and high-performance sports cars. Thus, in the mid-20th century, engineers and designers began to recognize the potential of rubber beyond tires. It was quickly integrated into various car parts to improve performance, safety, and comfort.

One of the key areas where rubber made a big impact was vibration control systems. Engine mounts made primarily of rubber were introduced to absorb and dampen the vibrations produced by the engine [7]. This reduced noise and discomfort inside the vehicle and minimized wear and tear on the engine and chassis. This application showed how the material can improve the driving experience and extend the vehicle's life.

In addition to the engine, rubber also found its way into suspension systems. The rubber bushings and dampers have been designed to absorb road shock and promote a smoother ride [8]. The flexibility of the rubber was ideal for these applications and provided the necessary resistance and durability under constant traffic conditions.

The rubber was further used to manufacture various automotive seals and seals throughout the vehicle [9]. These rubber components played an important role in preventing leaks, reducing internal road noise, and protecting sensitive parts from the elements. From the seals around the doors and windows that kept out water and wind to the seals in the engine that helped keep oils and gases out, rubber kept almost every part of the car working properly.

Before the development of synthetic alternatives, these components were historically manufactured from natural rubber, which exhibited high tensile strength and elasticity [10, 11]. As the most readily available and practical material for various automotive components, the properties of natural rubber were ideal for the less complex rubber components associated with the first automobiles. Natural rubber was also relatively easy to work into various shapes and forms, providing versatility. The trajectory of organic/natural fibers employed in the automobile sector is shown in Fig. 11.3.

Advances in rubber technology have significantly impacted the automotive industry, allowing for more durable products with improved performance. Over the decades, these advances have led to more complex rubber compounds and manufacturing technologies that can be tailored to the specific requirements of automotive applications.

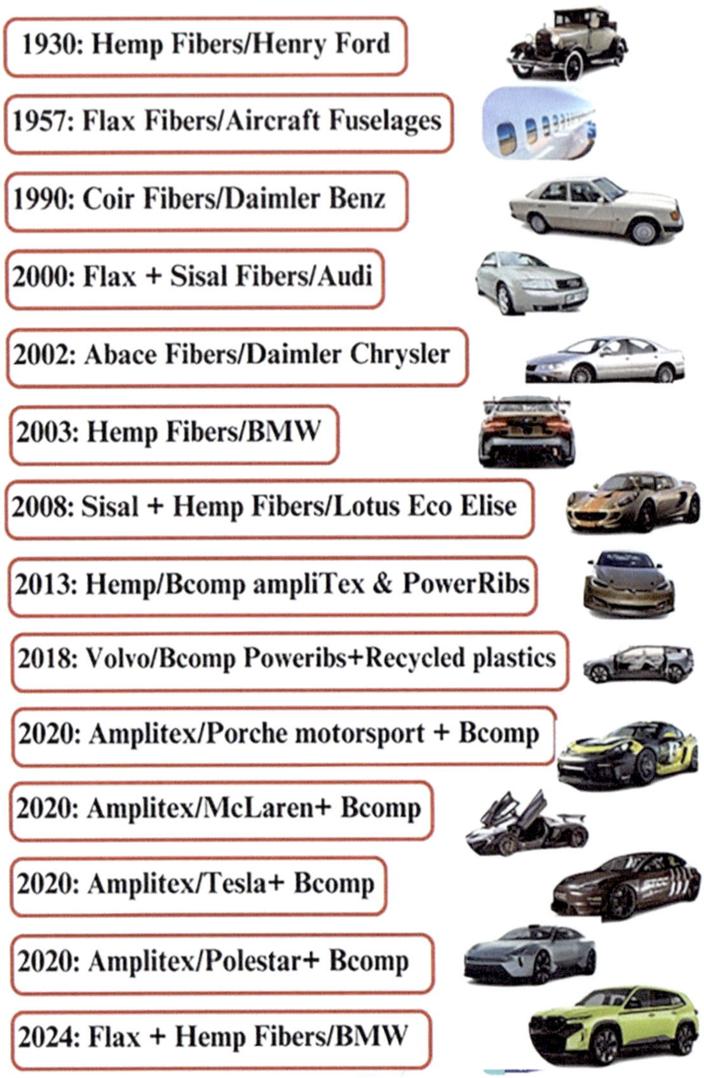

Fig. 11.3 Timelines for using natural fibers in the automobile sector [5]

One of the main breakthroughs was the creation of synthetic rubber [12]. This innovation was created in response to the limitations of natural rubber, including its availability and performance in extreme conditions. Synthetic rubbers such as styrene-butadiene rubber (SBR) and ethylene-propylene-diene monomer (EPDM) [13] provided improved resistance to abrasion, heat, and aging. This was particularly useful in the tire industry, where these properties provided longer tire life and better performance.

Further progress has been made in formulating rubber compounds using carbon black and silica additives [14]. Manufacturers managed to significantly improve the properties of rubber to improve tensile strength, heat resistance and elasticity. This made it possible to expand the range of automotive applications.

EPDM is widely used to produce most extruded rubber products for the automotive industry [13]. Because EPDM has excellent resistance to weathering, UV radiation, ozone, and temperature changes, it can be used around doors and windows without the risk of deterioration. Thanks to its resistance to heat and chemicals, it can also be used for under-the-hood applications.

In the 21st century, there is an ongoing emphasis on sustainability, including bio-rubber research and recycling processes [1]. The use of renewable rubber materials in automotive applications is part of a wider industry trend towards sustainability. It aligns with the growing consumer demand for environmentally friendly products and the industry's commitment to reducing its carbon footprint and dependence on non-renewable resources.

11.2 Rubber Materials

Rubber products that meet various requirements for heat resistance, frost resistance, chemical resistance, hardness, elasticity, and flexibility are widely used in automobiles [15]. The starting material for rubber is natural or synthetic rubber (Table 11.1). The isopentene groups of polyisoprene are linked into a long molecular chain, which allows it to be classified as a linear polymer with a carbon chain.

Polyisoprene and its analogs [16], as well as polybutadiene and polychloroprene, belong to the class of elastomers. Elastomers are polymers with high elasticity over a wide range of temperatures. They can undergo significant (up to a thousand percent or more) reversible deformations with a modulus from hundreds to several kilograms per 1 cm^2. This ability is manifested in the interval between the crystallization or glass transition temperature and the temperature at which the elastomer passes into a viscous state.

Elastomers include copolymers of diene and vinyl derivatives (butadiene-styrene, butadiene-nitrile, butyl rubbers, copolymers of ethylene and polypropylene, etc.). Many elastomers belong to organoelement compounds, for example organosilicon polymers and fluorine-containing polymers. A common feature of elastomers is their macromolecular structure and ability to vulcanize. In their original state, rubbers (elastomers) are not used for parts due to their high ductility. The transition of plastic rubber to elastic is called vulcanization, the cross-linking of linear macromolecules into a liquid network structure [17]. Sulfur, thiuram, diphenyl guanidine, and diazo compounds are vulcanizing agents and accelerators. The composition of rubber compounds (raw rubber) includes, in addition to rubber and vulcanizing agents, fillers (chalk, talc, zinc oxide, silicon dioxide, magnesium carbonate, carbon black), stabilizers, plasticizers, vulcanization retarders, and other special additives

Table 11.1 Characteristics of rubbers and their purpose

Names of rubbers	Glass transition temperature, °C	Purpose
Isoprene rubbers		
Natural	−70 … −72	General purpose, tires, and rubber engineering products
Synthetic	−66 … −70	General-purpose tires, shock absorbers, spongy products
Butadiene rubbers	−95 … −110	Protective, covering rubbers (frame, breaker sidewall)
Styrene-butadiene rubber (SBR)	−50 … −65	General purpose, tires, and rubber engineering products
Ethylene glycol rubbers	−55 … −65	Sponge products, car window gaskets, and other car parts
Butyl rubber	−67 … −69	General and special purpose, car cameras, rubberized fabrics
Nitrile butadiene rubber (NBR)	−25 … −50	Oil- and gasoline-resistant rubber products (gaskets, cuffs)
Chloroprene rubbers	−40	V-belts, gaskets and sealing parts, outer layer of tire sidewalls
Fluorine rubbers	−20 … −40	Sealing of products working in contact with oils, fuels, and aggressive liquids
Siloxane rubbers	−78 … −130	Sealing and vibration-isolating gaskets

[18]. In general, more than 350 additives are used in producing rubber compounds in addition to rubber. The vulcanized rubber compound is called vulcanizate or rubber.

A partial substitute for rubber is regenerated plastic, obtained by processing old rubber products (tires, inner tubes, galoshes, etc.) and waste from rubber production [19]. Rubber products with regenerated material have increased resistance to oils and gasoline and resistance to acids and bases, but at the same time, their physical and mechanical properties are reduced.

Depending on their purpose and operating conditions, rubber compounds are divided into groups for which numerical symbols are set for the temperature range of rubber properties. For example, group rubber with the symbol I is operational from −10 to +100 °C. In some cases, the maximum temperature for the use of rubber can be increased. In the environment of superheated water, rubbers of group U type 51–1481 and IRP-1375 based on ethylene propylene rubber are operable up to 170 °C. The temperature range for rubber use may be extended based on product performance test results.

11.3 Properties of Rubber Products

Rubber products and materials are classified according to the following criteria: purpose, conditions of use (temperature, environment, pressure, etc.), design (rubber, rubber-metal, rubber textile, monolithic, sponge, hollow, etc.), material (rubber classes and types of rubber, brands reinforcing materials), production technologies [20, 21].

The main properties of rubber as construction material include [15]:

- low modulus values in shear, tension, and pressure;
- great influence of the duration of the applied load and the temperature factor on the dependence of stress on deformation;
- almost constant volume during deformation;
- significant mechanical losses during cyclic deformations.

Under the influence of various factors of storage and operation, acting separately or most often in combination, rubbers change their valuable technical properties: elasticity decreases, hardening occurs, brittleness occurs, cracks appear, and color changes. Exposure to oxygen, especially ozone, causes the rubber to age and fatigue [22]. This is facilitated by heat and light, stress from static or dynamic loads, irrational storage, an aggressive environment, or the catalytic effect of metal salts (especially natural rubber is affected by manganese and copper salts). Low temperatures lead to a decrease in rubber elasticity and brittleness. These changes for stressed rubbers based on crystallizing rubbers increase with increasing cooling time. However, when moving to room temperature, the original properties are restored.

11.4 Application

11.4.1 Cuff Radial Seals

Various sealing products using rubber materials are most important in the automotive industry. Depending on the type of joint to be sealed, a distinction is made between the sealing of fixed joints, where the contact surfaces do not move relative to each other, and the sealing of joints, the contact surfaces of which carry out mutual reciprocating or rotational movement. movement.

Rotary motion hinge seals are designed to create a seal (tightness) where axle shafts and other moving elements exit the internal cavities of the units. The simplest, cheapest, small-sized types of seals for hinges with rotary movement are rubber-reinforced cuffs—rubber-reinforced cuffs with one edge with a spring for sealing shafts working in mineral oils, water, diesel at overpressure up to 0.05 MPa, movement speed up to 20 m/s and temperatures from −60 to +170 °C depending on the rubber group.

Rubber parts for sealing the fixed and movable joints of the hydraulic drive of the braking system and clutch units of the car, designed to work in contact with brake fluid at a pressure of up to 14.7 MPa, as well as to protect the units from moisture and dust [23].

11.4.2 Vibration Isolation Rubber Products

The group of vibration-isolating parts includes engine and transmission mounts, cabs, torque couplings, shock absorbers, compression bumpers, etc. The main purpose of rubber vibration isolation products is to provide protection against vibrations and quickly dampen vibrations in various components and assemblies of machines and engines. Rubber-metal products are used in machines, which is due to the need for their mechanical fastening in components and assemblies.

Vibrating rubber products must meet the following general requirements:

- long-term performance under various deformations (tension, pressure, shear, torsion, as well as several types of deformations at the same time—compression with shear, compression with torsion);
- low residual deformations, no more than 10%;
- high strength of the connection between rubber and metal.

For each specific group, certain operational requirements must be met by their purpose. For example, engine mounting cushions sense the power unit's weight during operation, limit its movement, and reduce dynamic loads, noise, and vibration transmitted through the cushions to the machine's chassis.

The requirements for vibration-isolating materials are best met by rubber based on natural rubber, which has high elasticity, resistance to long-term and repeated deformations in a wide frequency range, and satisfactory damping properties.

Recently, synthetic rubbers based on isoprene rubber, which have somewhat less modularity and resistance to aging than natural rubbers, have been widely used to produce rubber products.

The vibrating parts work at 60…130 °C depending on the rubber band. The percentage source of gamma for vibration isolators is 80%, and for cases made of rubber-metal compounds—90%.

11.4.3 Sealing of Brake Systems

Hydraulic brake cuffs are installed in friction units, ensuring machine control and operational safety. Depending on the operating conditions, such cuffs have different designs: cup, cellular, and combined rings of round and rectangular parts.

Rubber sealing and protective parts for movable joints of the hydraulic drive of brake system assemblies and clutches are designed to work in contact with brake fluid up to a pressure of 14.7 MPa, as well as to protect assemblies from dust and moisture, made of oil-resistant rubber, usable at temperatures −60 °C… +200°C.

Gasket design and sizes depend on the configuration of the gasket gap. The rubber of these products must be long-term resistant to aggressive weather effects during the operation of machines in all climate zones. Depending on the rubber brand, the service life in gamma percentages is: for monolithic seals—from 3.5 to 8.6 years and for sponge seals—from 1 to 6 years.

11.4.4 Hoses

Hoses are used as flexible pipes that ensure the transport of liquid media (fuel, oil, etc.) to the components of various machine systems. Very high demands are placed on the sleeves. During prolonged operation under pressure, elevated temperatures, and in an aggressive environment, they must maintain strength, tightness, and flexibility, long-term resistance to the effects of the environment, withstand external mechanical loads, and maintain dimensional stability.

In machines and engines, hoses reinforced with cotton and synthetic fibers and braided construction hoses work in fuel, brake fluid, water, and antifreeze environments.

Special requirements are placed on brake hoses because the reliability of the vehicle's brake system and, as a result, safety depends on their reliability and efficiency.

Curved cooler hoses with an internal diameter of up to 30 mm, through which the coolant is supplied under a pressure of up to 0.2 MPa, must be sealed with a residual pressure of 0.45 MPa. With an internal diameter over 30 mm, they must withstand a pressure of 0.35 MPa. The ends of the hoses must withstand a radial extension of up to 10% of the internal diameter.

11.4.5 Liquid Gaskets

Among the many sealing options in modern machines, flange joints have been expanded using solid seals made of cardboard, Maronite, and other polymeric materials.

The disadvantages of such connections include the need to create a certain pressure on the gasket because in the process of residual deformations in the gasket, the tightness of the joint is disturbed; with long-term use (or storage), the sealing materials age, which also leads to a violation of the tightness of the connection; the need for precise adjustment of the contact surfaces (distortions and deformations are not allowed), as well as a strict sequence when tightening the threaded connections when creating pressure on the seal. All this reduces the reliability of the flange connection and the assembly.

A new direction in the field of joint sealing is using viscous materials for sealing flange joints, called liquid seals. They are divided into drying, non-drying, and vulcanizing. Freely changing shape under force, filling micro- and macro-uniformities on the surface of the parts and creating a high degree of tightness of the joints; the presence of adhesion between the liquid sealing material and the part increases the reliability of sealing the connection.

11.5 Summary

Rubber has been a key material in the automotive industry for over a century. Since its introduction to the market in the late 19th century with the creation of the tire, rubber has played a fundamental role in developing and improving vehicle technology, safety, and performance. Originally utilized primarily for tires, the use of rubber quickly expanded to encompass vibration control systems, suspension parts, and various seals, all aimed at improving vehicle performance, safety, and comfort. As time has progressed, the introduction of synthetic rubbers has further propelled advancements in the industry by enhancing resistance to wear, heat, and aging. Additionally, the automotive sector has investigated bio-rubber and recycling initiatives to promote sustainability.

Rubber materials are classified based on their properties (elasticity, resistance to extreme temperatures), and chemical durability. Both natural and synthetic rubbers are used. Rubber's flexibility and adaptability are further enhanced through vulcanization, a process that strengthens the material by cross-linking polymer chains. Fillers and additives, such as carbon black and silica, are also used to optimize mechanical properties and extend rubber's durability in automotive applications.

The automotive industry makes extensive use of rubber in a variety of components, such as radial seals for joints, vibration isolation mounts, seals for brake systems, hoses for fluid transfer, and liquid gaskets for flange connections. These rubber components play vital roles by providing sealing, noise reduction, vibration damping, and protection against environmental elements. Ongoing advancements in rubber technology are enhancing these applications, leading to better vehicle performance and durability, while also tackling sustainability issues through the use of recycled and renewable materials.

References

1. Utrera-Barrios S, Verdejo R, López-Manchado MA, Santana MH (2023) Self-healing elastomers: a sustainable solution for automotive applications. Eur Polymer J 190:112023. https://doi.org/10.1016/j.eurpolymj.2023.112023
2. Rubber parts for the automotive industry. https://www.rpmrubberparts.com/industries/automotive-industry/. Accessed 28 Feb 2025

3. Fimrite DE, McAfee CD (2021) 8 – Make rubber, sell rubber: building a sustainable industry. In: Letcher TM, Shulman VL, Amirkhanian S (eds) Tire waste and recycling. Academic, Elsevier, pp 145–163. https://doi.org/10.1016/B978-0-12-820685-0.00009-0
4. Ullah S, Akhter Z, Palevicius A, Janusas G (2025) Review: natural fiber-based biocomposites for potential advanced automotive applications. J Eng Fibers Fabr. https://doi.org/10.1177/15589250241311468
5. Vijayaram TR (2009) A technical review on rubber. Int J Des Manuf Technol 3(1):25–37. https://doi.org/10.18000/ijodam.70043
6. John Dunlop pneumatic tires. https://lemelson.mit.edu/resources/john-dunlop. Accessed 18 Jan 2025
7. Santhosh S, Velmurugan V, Paramasivam V, Thanikaikarasan S (2020) Experimental investigation and comparative analysis of rubber engine mount vibration and noise characteristics. Mater Today Proc 21(Part 1):638–642. https://doi.org/10.1016/j.matpr.2019.06.730
8. Huang Y, Zhang H, Fan Q, Huang Q, Shao L, Zhan X, Wang J (2024) The stiffness and damping characteristics of a rubber-based SMA composite shock absorber with a hyper-elastic SMA-constitutive model considering the loading rate. Materials 17:4076. https://doi.org/10.3390/ma17164076
9. Rubber seals in the automotive industry (2021). https://www.casertainc.com/blog/rubber-seals-automotive-industry?utm_source=chatgpt.com. Accessed 19 Jan 2025
10. Alarifi IM (2023) A comprehensive review on advancements of elastomers for engineering applications. Adv Ind Eng Polym Res 6(4):451–464. https://doi.org/10.1016/j.aiepr.2023.05.001
11. Elfaleh I, Abbassi F, Habibi M, Ahmad F, Guedri M, Nasri M, Garnier C (2023) A comprehensive review of natural fibers and their composites: an eco-friendly alternative to conventional materials. Results Eng 19:101271. https://doi.org/10.1016/j.rineng.2023.101271
12. Kabakçi GC, Aslan O, Bayraktar E (2022) A review on analysis of reinforced recycled rubber composites. J Compos Sci 6:225. https://doi.org/10.3390/jcs6080225
13. Costa NL, Hiranobe CT, Cardim HP et al (2024) A review of EPDM (ethylene propylene diene monomer) rubber-based nanocomposites: properties and progress. Polymers 16:1720. https://doi.org/10.3390/polym16121720
14. Robertson CG, Hardman NJ (2021) Nature of carbon black reinforcement of rubber: perspective on the original polymer nanocomposite. Polymers 13:538. https://doi.org/10.3390/polym13040538
15. De SK, White JR (2001) Rubber technologist's handbook. Rapra Technology Limited, UK.
16. Cruz-Morales JA, Gutiérrez-Flores C, Zárate-Saldaña D, Burelo M, García-Ortega H, Gutiérrez S (2023) Synthetic polyisoprene rubber as a mimic of natural rubber: recent advances on synthesis, nanocomposites, and applications. Polymers 15:4074. https://doi.org/10.3390/polym15204074
17. Sullivan AB, Wise RW (1999) Processing and vulcanization tests. In: Morton M (ed) Rubber technology. Springer, Dordrecht. https://doi.org/10.1007/978-94-017-2925-3_4
18. Vayyaprontavida Kaliyathan A, Varghese K, Nair AS, Thomas S (2019) Rubber–rubber blends: a critical review. Prog Rubber Plast Recycl Technol 36(3):196–242. https://doi.org/10.1177/1477760619895002
19. Lapkovskis V, Mironovs V, Kasperovich A, Myadelets V, Goljandin D (2020) Crumb rubber as a secondary raw material from waste rubber: a short review of end-of-life mechanical processing methods. Recycling 5:32. https://doi.org/10.3390/recycling5040032
20. The evolution of rubber in automotive manufacturing (2024). https://interplasinsights.com/plastic-industry-insights/latest-plastics-industry-insights/the-evolution-of-rubber-in-automotive-manufacturing/#:~:text=Rubber%20has%20been%20a%20key,vehicle%20technology%2C%20safety%20and%20performance. Accessed 18 Jan 2025
21. Bilichenko VV (2025) Materials for car service and repair. Rubber and sealing products. https://web.posibnyky.vntu.edu.ua/fmbt/bilichenko_servis_ta_remont_avto/2.1.html. Accessed 18 Jan 2025

22. Rezig N, Bellahcene T, Aberkane M et al (2020) Thermo-oxidative ageing of a SBR rubber: effects on mechanical and chemical properties. J Polym Res 27:339. https://doi.org/10.1007/s10965-020-02330-y
23. Automobile industry. https://www.industrialrubbergoods.com/automobile-industry.html. Accessed 18 Jan 2025

Chapter 12
Smart Materials

Abstract This chapter explores the transformative role of smart materials in the automotive industry, emphasizing their capacity to enhance safety, efficiency, and adaptability. It details key types of smart materials, including piezoelectric materials, shape memory alloys (SMAs), thermoelectric materials, self-healing polymers, magnetorheological fluids, electrochromic glass, biomimetic materials, and nanomaterials. Piezoelectric materials are highlighted for their applications in sensors, actuators, and energy harvesting systems, with classifications including ceramics, polymers, and composites. SMAs, particularly Ni–Ti and copper-based alloys, offer self-adjusting functions through thermomechanical phase transitions and are used in actuators, injectors, and adaptive systems. Integrating SMA-based composites enhances energy absorption, stiffness, and damage tolerance. These smart materials contribute to weight reduction, fuel efficiency, and sustainability. The chapter concludes that ongoing smart materials innovation will further enable intelligent automotive systems development, support environmental goals, and advance user experience.

12.1 Introduction

The modern generation of cars must meet a range of safety, comfort, and performance characteristics in an increasingly competitive and progressive automotive industry. Therefore, intelligent materials with unique and exceptional properties are gaining momentum [1]. Smart materials can change shape, position, elasticity, and fluid flow rate. They can quickly react to changes in temperature and electric and magnetic fields [2].

Automotive smart materials market size was valued at USD 5.10 Billion in 2023 and is expected to reach USD 12.10 billion by the end of 2030 with a CAGR of 6.83% during the forecast period 2023–2030 [3].

The market for automotive smart materials is expanding due to several important considerations. The demand for these cutting-edge materials is being driven

K. Berladir et al., *Materials for the Automotive Industry*, Springer Tracts in Mechanical Engineering, https://doi.org/10.1007/978-3-032-05866-9_12

by the growing emphasis on energy efficiency, reducing vehicle weight, and developing environmentally friendly and sustainable transportation options. The pursuit of improved safety, comfort, and aesthetics in automobile design and the incorporation of cutting-edge technology technologies in cars also fuels the market's growth.

Ultimately, the automotive smart materials market will play a major role in transforming the functioning and design of automobiles and launching a new wave of intelligent, adaptable, and creative automotive solutions that will enhance performance, safety, and environmental sustainability [4].

The classification of smart materials is presented in Fig. 12.1.

The modern smart materials used in the automotive industry include [1–7]:

- **Piezoelectric materials** can generate electricity during a mechanical impact or vice versa. They can be used to create smart sensors and actuators that monitor and control various aspects of the vehicle, such as tire pressure, engine vibration, or braking force. Piezoelectric materials can also be used to harvest energy from vehicle motion, such as suspension, exhaust, or wind, and use it to power other devices or charge a battery [8].
- **Shape memory alloys (SMAs)** are metals that can return to their original shape after being deformed by heat or electricity. They can be used to create adaptive components that adapt to different driving conditions, such as aerodynamic spoilers, adaptive suspension systems, or self-healing bumpers. SMAs can also

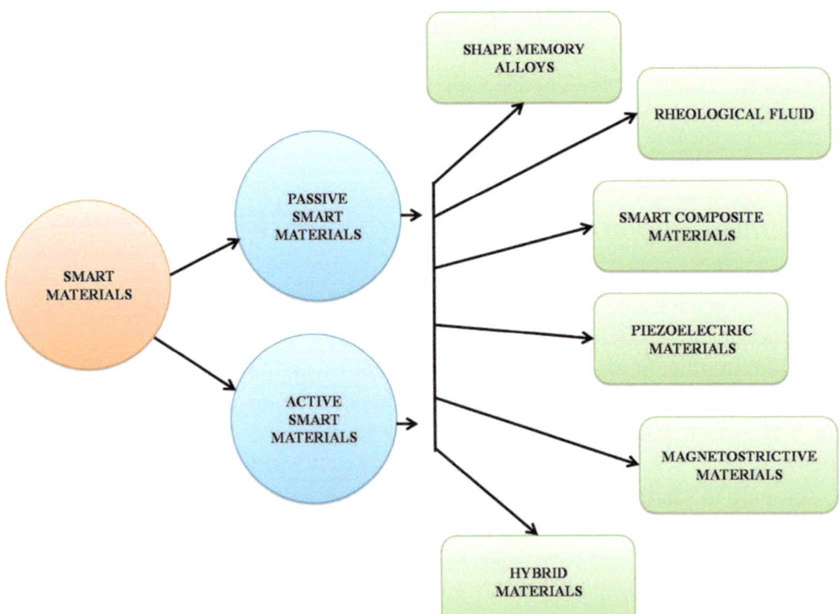

Fig. 12.1 Classification of smart materials [5]

reduce weight and complexity by replacing conventional mechanical or hydraulic systems with simpler, lighter designs [9].

- **Thermoelectric materials** can convert heat into electricity or vice versa. They can be used to create intelligent cooling and heating systems that regulate the temperature inside and outside the vehicle, improving comfort and performance. Thermoelectric materials can also be used to recover waste heat from engines or brakes and use it to generate electricity or power other systems. Thermoelectric materials can also reduce emissions and fuel consumption as they can optimize a vehicle's thermal efficiency [10].

- **Self-healing materials** can repair themselves after being damaged by external factors such as scratches, cracks, or corrosion. They can be used to create smart coatings and structures that protect the vehicle from wear and tear, reducing maintenance and replacement costs. Self-healing materials can also improve safety and reliability because they can prevent or minimize the impact of structural failures or malfunctions. Self-healing materials can also improve the appearance and durability of a car, as they can restore its original color and shine [11–15].
- **Magnetostrictive materials** convert electromagnetic energy into mechanical energy and vice versa. This effect can be used to create sensors that measure a magnetic field or detect a force. The magnetic field or force would create a strain in the material, which can be measured [16].
- **Magnetorheological fluids** are intelligent materials whose rheological and mechanical properties can be modified by interaction with an external magnetic field. These unique features allow for a controlled change of their viscosity, which is applied in technology to build adaptive devices and effectively suppress vibrations in various mechanical systems [17].
- **Electrochromic glass** can change its transparency or color by applying a small voltage. It can be used to create smart windows that regulate the amount of light and heat entering the car, improving comfort and energy efficiency. Electrochromic glass can also provide privacy and security, as it can block the outside view or display information or images on the surface of the glass. Electrochromic glass can also enhance the aesthetic appeal of a vehicle, as it can create dynamic and customizable color effects [18].
- **Biomimetic materials** mimic the properties or functions of natural materials or organisms [19]. They can be used to create smart features that adapt to the environment or driver preferences, such as camouflage, ventilation, or communication. Biomimetic materials can also inspire new, more efficient, sustainable, or innovative design solutions, such as lightweight structures, renewable energy, or biodegradable components [20].
- **Nanomaterials** [21].

12.2 Piezoelectric Materials

Piezoelectric materials are smart materials that convert mechanical energy into electrical energy by applying stress or strain and vice versa. Automotive technology is currently the second-largest market for piezoelectric products. Piezoceramics are used in the automotive industry to produce high-precision sensors for a variety of applications such as engine control, parking assistance, level and acceleration sensors, and tire pressure monitoring. Piezoceramics provide precise measurement and control to improve vehicle performance and safety [22].

The nature of piezoelectric materials is closely related to the significant number of electric dipoles in these materials. These dipoles can be caused by ions at lattice sites with an asymmetric charge environment or by certain molecular groups with electrical properties. A dipole is a vector with direction and magnitude according to the electric charges around it. These dipoles usually have the same direction when next to each other, and their combination forms regions called Weiss domains. Normally, the domains are randomly oriented, but they can be oriented by polarization, a process in which a strong electric field is applied to the material. Piezoelectric material generates voltage because the crystal structure changes under mechanical influence, and thus, the direction of polarization of electric dipoles changes. Depending on the nature of the dipole, this change in polarization can be caused by either the reconfiguration of ions in the crystal structure or the reorientation of molecular groups. Therefore, the greater the mechanical stress, the stronger the polarization change and the more electricity is generated. A common piezoelectric ceramic is a mass of perovskite ceramic crystals, each consisting of a small tetravalent metal ion, typically titanium, in a lattice of larger divalent metal ions, typically lead and oxygen. Many crystals can have dipole moments under conditions that give tetragonal or rhombohedral crystal symmetry. The change in polarization manifests itself as a change in the surface charge density on the crystal surfaces, i.e. as a change in the electric field between the surfaces.

12.2.1 *Types of Piezoelectric Materials*

Piezoelectric materials can be classified into three main categories (Fig. 12.2).

Ceramic piezoelectric materials are usually polycrystalline materials with a perovskite crystal structure. These materials show significant anisotropy, due to which their physical properties change depending on the direction of the applied electric field.

A well-known piezoelectric material is quartz. Due to its crystal structure, it exhibits strong piezoelectric energy, which means that when pressure is applied to the quartz crystal, electrical polarization can be observed in the direction of pressure. Gallium orthophosphate has almost the same crystal structure as quartz, so it has the same properties. However, its piezoelectric effect is almost double that of quartz,

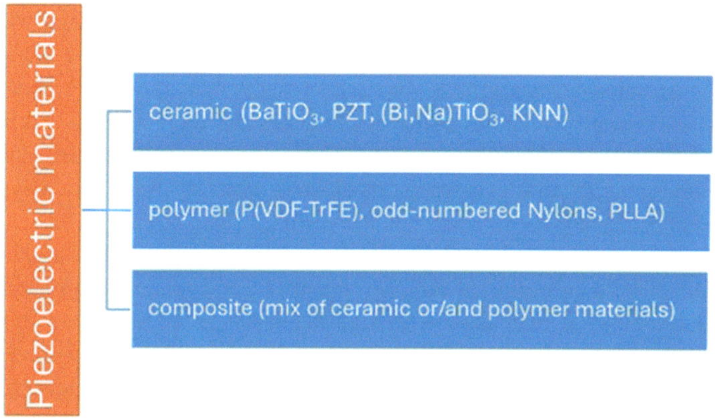

Fig. 12.2 Classification of the piezoelectric materials

making it valuable for mechanical applications. It is not a natural element; it is necessary to synthesize it. Tourmaline crystal is usually black but can vary from purple to green and pink.

These materials are characterized by exceptional piezoelectric properties, and lead zirconate titanate (PZT) is considered one of the most important piezoceramic materials widely used in various fields of application. Other examples of ceramics include potassium niobate ($KNbO_3$) and barium titanate ($BaTiO_3$). Barium titanate is an electroceramic that can replace lead zirconate titanate for piezoelectric energy. It is used for microphones and transducers. It is considered one of the economical piezo elements. Therefore, it is used in many applications.

Polymer piezoelectric materials are flexible materials that can undergo significant mechanical deformation while maintaining their piezoelectric effect. They are also light and economical, suitable for low-cost and low-temperature production processes. Due to their flexibility, polymeric piezoelectric materials can be easily shaped to specific requirements, making them attractive for various applications.

The study of piezoelectric polymers has a long history from the beginning of the 20th century. However, the strong piezoelectric effect in polar polymers was first discovered by Kawai's work in 1969, in which a polyvinylidene fluoride (PVDF) film was stretched and polarized. Since then, PVDF and its copolymers have been widely researched and developed for various applications. They are considered the most important piezoelectric polymers, and research continues to improve their performance while taking sustainability issues into account.

Composite piezoelectric materials are smart materials that combine ceramics or polymers with other materials to exploit their respective properties. Piezoelectric composites typically consist of a piezoelectric ceramic filler embedded in a piezoelectric polymer matrix, and the properties of the composite can be tailored to specific applications by adjusting the ratio of the materials. These materials provide several advantages, including low acoustic impedance, high coupling coefficients, mechanical flexibility, favorable compatibility with human tissue, and wide bandwidth combined with a low mechanical quality factor. As a result, piezoelectric composites are suitable for applications such as medical diagnostic ultrasound transducers and underwater sonars. Regarding the harmful effects of lead, it is important to minimize the release of lead into the environment to protect the health of the planet and its inhabitants. Several lead-free piezoelectric materials have been developed in recent years.

Piezoelectric materials based on alkali niobate. Niobate-based alkaline materials consist of solid solutions of $KNbO_3$, $NaNbO_3$, and related compounds. These materials have evolved over time with significant advances in understanding their electrical properties and physical mechanisms. Matthias discovered the ferroelectric properties of $KNbO_3$ materials, while Woosden proposed an antiferroelectric model for $NaNbO_3$. The binary system $KNbO_3$–$NaNbO_3$ (KNN) was first studied by Sheeran et al. The different phase transition temperatures for these materials and the study of phase boundaries are currently a major area of research.

$Bi_{0.5}Na_{0.5}TiO_3$-based piezoelectric materials. The compound Bismuth sodium titanate (BNT, $Bi_{0.5}Na_{0.5}TiO_3$) was first discovered as a rhombohedral structure by Smolensky et al. Since then, BNT materials have attracted attention due to their high remanence polarization. However, it is difficult to fabricate pure BNT ceramics with desired piezoelectric properties due to their high conductivity and large coercive field, resulting in poor piezoelectric energy. Consequently, chemical modification has become a common approach to improve the piezoelectric behavior of BNT materials. Considerable progress has been made in the area of piezoelectric and deformation properties. However, the depolarization temperature limits the practical use of these materials in piezoelectric devices [23].

$BaTiO_3$-based piezoelectric materials. Barium titanate ($BaTiO_3$, BT) is the first perovskite to produce lead-free piezoelectric ceramics. However, as observed, the conventional solid-state response method only resulted in a weak piezoelectric effect for many years. However, progress has been made in this material. In 2009, Ren et al. discovered a significant piezoelectric effect (d33.620 pC/N) in a lead-free $Ba(Ti_{0.8}Zr_{0.2})O_3$-$(Ba_{0.7}Ca_{0.3})TiO_3$ (BCT-BZT) ceramic with a. new phase boundary. In addition, new properties, such as the electrocaloric effect, have also been discovered in BT-based materials.

12.2.2 Application

Piezoelectric materials are used in the automotive industry in two ways: directly and indirectly. Both have significantly impacted the production and operation of automobiles, but their applications are completely different [24–28].

Direct application piezoelectric materials are integrated into vehicles to achieve various functions. These can be actuators, injectors, and sensors that control, regulate, and read various systems. Piezoelectric materials produce automotive devices that can achieve precise and reliable control over billions of cycles and are repeatable and accurate. This is a huge advantage for automotive applications, especially in fuel economy, performance, and safety systems, where malfunctions can cause serious problems.

Indirect use of piezoelectric materials in the automotive industry includes their use in designing, manufacturing, and testing machines and robotic vehicles. The high precision and reliability of such materials used in various transducers, sensors, and actuators make them ideal for indirect automotive applications. Piezoelectric technology plays a significant role in the automotive industry, from devices used to adjust headlights to the precise positioning of mechanical arms used to install components.

Pressure sensors. Primarily used to measure pressure and force, they can also be used to measure internal combustion pressure and vacuum in engines. They can measure large and small pressures, such as intake manifold MAP and knock sensors. Because the piezoelectric pressure sensor's electric charge after applying an external force can only be preserved when the circuit has an infinitely large input resistance, it can only be used for dynamic voltage measurement and not for static measurement. Due to its ability to withstand high temperatures and high humidity, it is widely used in the automotive industry and other fields.

Knock sensors. The knock sensor comprises a piezoelectric generator, a metal sheet, a sealed rod, a metal cover, etc. The electric charge generated by the piezoelectric generator is proportional to the vibrations of the engine cylinder, and the generated voltage is supplied to the electronic control unit through a shielded conductor. The electronic control unit determines the detonation intensity based on the vibration voltage's magnitude of around 7 kHz. It promptly adjusts the ignition timing, allowing the engine to operate at the ignition point closest to maximum thermal efficiency and minimum fuel consumption. Operating condition without detonation and guaranteeing the engine's operation with the highest possible level of power and efficiency.

Piezoelectric actuators. They play an important role in many different vehicle components. The drive converts electrical signals into a precise physical movement called displacement. This shift can fine-tune lenses, reflectors, and other components, act as a low-volume pump, or control a hydraulic valve. Piezoelectric automotive actuators can be found in any application that requires simplicity and reliability. Piezoelectric actuators require no maintenance and can operate billions of times without wear or degradation.

Piezoelectric injector. Fuel injectors using piezoelectric technology are more precise than their conventional counterparts. Using piezoelectric elements, the hinge in the fuel injector can open and close faster than traditional fuel injectors, allowing for more precise control of fuel injection. Fuel injectors using piezoelectric technology have been proven to improve fuel consumption and reduce emissions.

Ultrasonic sensors. They are backup devices for reverse collision, rear radar, or sonar, especially for large vehicles such as extended dump trucks, heavy trucks, mining vehicles, etc. Ultrasonic transducers typically comprise an aluminum alloy housing, piezoelectric ceramic transducers, sound-absorbing materials, and lead wires. They have wide horizontal directivity and limited vertical directivity. It uses the positive and negative piezoelectric effect of PZT piezoelectric ceramics to convert electrical energy into mechanical energy and vice versa. When an electrical signal is applied to the piezoelectric ceramic, mechanical movement is generated, and ultrasonic waves are emitted. When ultrasonic waves encounter obstacles while moving through the air, they are immediately reflected back. An electrical signal is sent when the reflected waves hit the piezoelectric ceramic. The distance between the vehicle and the obstacle is calculated and displayed by measuring the time difference through data processing. It can accurately detect small obstacles behind the vehicle and in the driver's blind spot and is very practical. To achieve high transmission efficiency and sensitivity the combined transmission and reception ultrasonic sensor is currently the main product in the market with high transmission efficiency, sensitivity, and clear directivity. Ultrasonic waves have a specific detection angle and range and can cover the car's rear. Ultrasonic sensors are also used in electronic suspension systems to directly monitor the distance between the vehicle chassis and the road surface.

Piezoelectric accelerometers. They are used in automotive airbag systems. They use the inertial force created by the inertia of the collision to create a proportional electric charge and voltage through the shear force acting on the piezoelectric ceramic body. They have high accuracy and good reliability. The accelerometer comprises two piezoelectric ceramic plates connected by an internal common electrode to form a two-stage structure, mounted in the direction of motion to form a cantilever beam, and integrated into a housing with a thick peripheral circumference. When it detects the intensity of vehicle collisions at low or high speed, it converts the signal into an electrical output signal. It meets the requirements of various diagnostic control algorithms, ensuring accurate and timely airbag opening at high collision intensity, increasing the vehicle's safety.

12.3 Shape Memory Alloys

Shape memory alloys (SMAs) are smart materials that can remember or retain their previous shape when exposed to external influences such as thermomechanical or magnetic gradients [29].

The automotive industry offers various applications for shape memory alloys, especially future developments. Although shape memory alloys are currently mainly used in the cooling circuit or to control mixing valves, they will play a crucial role in automotive automation in the future. Figure 12.3 overviews existing and potential SMA applications for passenger cars and classifies them by vehicle functional areas [30].

SMA is a group of metallic materials that can return to their original shape after being deformed by a certain force, such as mechanical stress, heat, or magnetic field. These alloys have two different crystal structures called austenitic and martensitic phases. The transformation between these phases is responsible for the shape memory effect and is reversible under certain conditions.

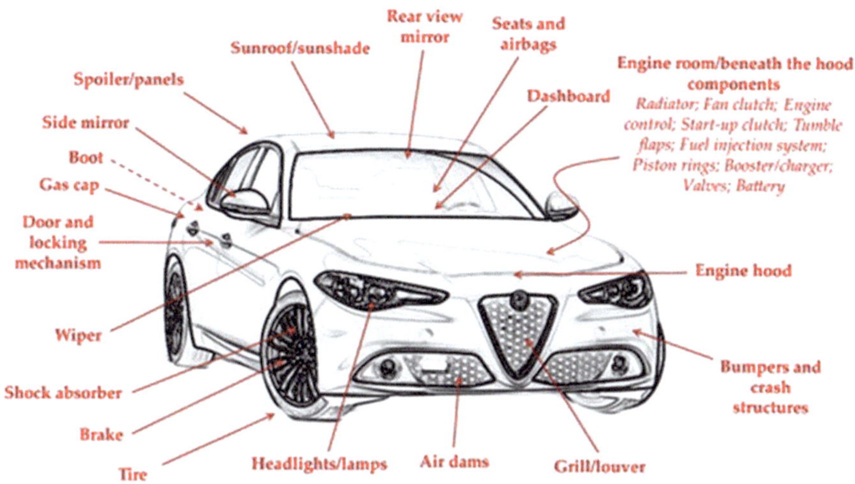

Fig. 12.3 Existing and potential shape memory alloy applications in the automotive field. *Source* [30]

Materials with a shape memory effect are characterized by a shape memory effect, superelasticity, pseudoelasticity, high damping properties, and adaptive properties associated with the reversibility of phase transitions.

The shape memory effect (SME) is a unique property of these materials, which allows them to remember their original shape even after several deformations. Shape changes due to temperature fluctuations are primarily associated with transforming the martensite phase (Fig. 12.4). SME is a phenomenon in which a material can regain its original shape even if it is deformed at low temperatures. The superelasticity (ES) effect is characterized by deformation recovery after unloading. This behavior is observed during loading and unloading at high temperatures and is attributed to a stress-induced martensitic transformation [31–33].

The mechanical behavior of shape memory alloys results from a transition between two different phases with defined crystal structures, namely the austenite and martensite phases. The austenite phase is stable at high temperatures and has a symmetrical structure. In contrast, the martensitic phase, which is stable at low temperatures, generally has an asymmetric structure. A drop in material temperature or increased stress causes a transition from the austenitic to the martensite phase. This can be considered a non-diffusive solid–solid transformation.

12.3.1 Types of Shape Memory Alloys

Ni–Ti shape memory alloys. Ni–Ti shape memory alloys are made using approximately equal amounts of nickel and titanium. In addition to these two elements,

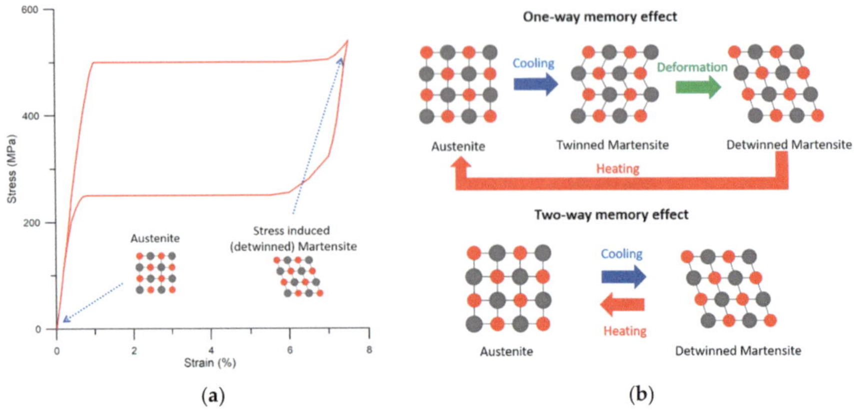

Fig. 12.4 Phase transformation in shape memory alloys [34]

copper, zinc, gold, and iron can be used to make shape memory alloys, but Ni–Ti is the most common memory alloy [35].

Nitinol (nickel-titanium or Ni–Ti) is an alloy of the equiatomic composition of titanium and nickel, where about 55% by weight is Ni, and the rest is Ti. The alloy was patented in 1965 by W. Bouguer and F. Wang of the Naval Weapons Laboratory.

Ni–Ti austenite has a cubic structure with relatively high entropy, which leads to energy dissatisfaction. When austenite is cooled, there is a significant decrease in Young's modulus. Upon further cooling, the austenite phase transforms into two alternative phases with lower entropy. These phases include B19 monoclinic and deformed orthorhombic martensite (R-phase) [36, 37].

Both depend on the thermodynamic and kinetic conditions of their formation. The martensitic phase is more thermally stable than the R phase and is therefore preferred. The R-phase forms earlier due to its low activation energy, giving it a kinetic advantage. Moreover, the martensitic crystal structure can be found either in a twinned form (stress-induced transformation) or a bifurcated form (temperature-induced transformation).

Some physical and mechanical properties of SMA differ between the two phases, such as Young's modulus, electrical resistance, thermal conductivity, and coefficient of thermal expansion. The austenite structure is relatively hard and has a much higher Young's modulus. In contrast, the martensite structure is softer and more ductile, meaning it can be easily deformed by external force (Table 12.1).

Reasons for using these two elements to create smart alloys include:

- Ni–Ti shape memory alloys are preferred because they are stable and practical and exhibit superior thermomechanical properties compared to other metals.
- The yield strength of Ni–Ti-based SMA alloys is lower than that of steel, but it is known that some compositions can have better yield strength than plastics and even aluminum. The yield strength of Ni–Ti alloys can reach 500 MPa.

Table 12.1 Commercial NiTi SMAs physical properties [38]

Property	Value	
	Martensite	Austenite
Corrosion resistance	Similar to 300 series SS or Ti-alloy	
Density ρ_D, kg/m^3	6450–6500	
Electrical resistivity (approx.) ρ_R, μΩ·cm	76–80	82–100
Specific heat capacity c, J/kg·K	836.8	836.8
Thermal conductivity k, W/m·K	8.6–10	18
Thermal expansion coefficient α, m/m·K^{-1}	$6.6 \cdot 10^{-6}$	$11.0 \cdot 10^{-6}$
Ultimate tensile strength σ_{ULT}, MPa	895 (fully annealed)/1900 (hardened)	
Young's modulus (approx.) E, GPa	28–41	75–83
Yield strength σ_Y, MPa	70–140	195–690
Poisson's ratio ν	0.33	
Magnetic susceptibility χ, μemu g	2.5	3.8

- The use of Ni–Ti-based alloys can lead to high levels of recoverable plastic deformation.

Copper-based shape memory alloys. Copper-based shape memory alloys are cheaper and easier to process than nickel-titanium memory alloys, so they have great potential for development [38]. However, the strength of copper-based shape memory alloys is not as good as that of nickel-titanium memory alloys, and the shape memory ability also degrades more quickly when reheated. Trace amounts of titanium, manganese, and zirconium can be added to improve the mechanical properties of copper-based memory alloys. Alloys of copper, zinc, and aluminum have the best performance characteristics and the widest application among copper-based shape memory alloys [39]. Copper, zinc, and aluminum alloys have high thermal conductivity and are sensitive to temperature changes. It can be made into temperature sensing elements and used as automatic control switches to regulate the temperature inside the car.

SMA-based composites. SMA has the potential to revolutionize the automotive industry by creating smart composite components that are more responsive, adaptable, and efficient than conventional composites in terms of weight reduction and energy absorption [40]. Several experimental and theoretical studies have been conducted on SMA hybrid composite for impact and energy absorption applications, some focusing on experimental studies and others on numerical simulations (Table 12.2).

In the automotive industry, advanced composite materials can enable intelligent component improvements. More specifically, SMA can be combined with other materials to create hybrid configurations that improve properties: stiffness, impact resistance, damping ability, durability, and multi-level interchangeability. These aspects will address the development of smart materials based on new concepts such as self-sensing, self-curing, self-healing, and self-protection methodologies to improve

Table 12.2 SMA-based composites for impact and energy absorption applications [23]

Matrix	Reinforcement	Results
Experimental studies		
BMI resin	CF/TiNi SMA trigger	SMA trigger provided a stable, progressive crushing process and a 37.2% increase in the specific energy absorption
Epoxy resin	GF/NiTi SMA wires	The embedment of SMA composites resulted in a 72.72% higher ballistic limit and better damage tolerance
Numerical studies		
Epoxy	CF/Nitinol SMA wires	Embedding SMA wires into composites leads to an increase in the damage resistance of hybrid composite structures
Epoxy	GF/NiTi SMA wires	Modeling of viscoelastic laminates showing good conformance with the experimental results
Epoxy	CF/Nitinol SMA strips	Embedding SMA strips were effective in improving the impact resistance of composite plates
Experimental and numerical studies		
PPS	CF/SMA wires	Embedding SMA wires increased the toughness of the composites and energy levels of absorbed energy before failure
Epoxy resin	GF/Nitinol SMA wires	Embedding SMA wires enhances the energy absorption capacity of the composite
Epoxy resin	GF/NiTi SMA wires	The SMA wires can improve damage tolerance and impact resistance of GFRP laminates

consumer safety, component life, and performance while reducing maintenance and manufacturing costs. Implementing SMA in primary structures can lead to significant compliance with industrial requirements and environmental policy, enabling the production of green and lightweight composites to reduce fuel consumption and air pollution. SMAs are also flexible in fabrication and can be engineered in various ways to produce new types (e.g., magnetic shape memory alloys—MSMA and HTSMA) or shapes (e.g., NiTi thin film) of SMAs. In addition, the new and improved SMAs have significantly improved the properties and performance of SMAs (e.g., higher stress and strain). These new types or forms of SMA have increased the potential use of SMA and surpassed its competitors [41].

12.3.2 Application

Recently, many automotive manufacturers are actively implementing SMAs in their vehicles. One of the earliest commercial SMA actuators for automotive applications is the thermally responsive pressure control valve embedded in Mercedez-Benz automatic transmissions for smooth gear shifting, first introduced in 1989. Later, Alfmeier Präzision AG (now Actuator Solutions GmbH) successfully mass-produced SMA pneumatic valves for lumbar support in car seats for Daimler Mercedes Benz, and today, this device is utilized by most automotive manufacturers such as BMW, General Motors (GM), Hyundai, Ford, Porsche, and Volkswagen (VW), with an annual production of 10 million actuators.

SMAs are excellent actuators for automotive applications due to their mechanical simplicity and better performance than most conventional actuators and are subsequently versatile to adapt to various design mechanisms and techniques.

Automotive actuators can be generally divided into three categories: (1) low-power actuators for comfort and bodywork functions, (2) high-power vehicle control actuators, and (3) high-frequency engine control actuators. SMA actuators are most suitable for the first category and practicable for the second category. However, SMAs are less preferred for the third category due to the low operational frequency and narrow bandwidth. Recently, newly developed SMAs, such as magnetic SMAs (MSMAs) and NiTi film SMAs, have increased the operational frequency up to 2 kHz, but it is still much slower than other conventional high-frequency actuators.

The combinations of SMAs with other materials to form shape-memory hybrids (SMHs) or shape memory composites (SMCs) have improved the material performance and functionality, such as active stiffening, higher damping capacity and toughness, and multiple-state changing capability. These new 'smart composites' are capable of doing multifunctional operations such as sensing and response simultaneously and possibly self-healing. SMAs are also flexible to be manufactured and designed in various ways to produce new types (e.g., MSMA and HTSMA) or forms (e.g., NiTi thin film) of SMAs. In addition, the new and improved SMAs have significantly enhanced SMAs attributes and performances (i.e. higher stress and strain). These new types or forms of SMAs have increased the potential of SMA utilization and outweigh their competitors [42].

12.4 Summary

Smart materials are transforming the automotive industry by introducing innovative solutions that enhance vehicle efficiency, safety, and adaptability. The ability of these materials to respond dynamically to environmental stimuli has paved the way for intelligent automotive components that improve performance and durability. Piezoelectric materials, for example, play a crucial role in sensors, actuators, and energy harvesting systems, enabling precise control and monitoring of various vehicle functions. Shape memory alloys offer unique benefits by enabling self-adjusting components, reducing mechanical complexity, and improving overall vehicle comfort and efficiency.

The continuous development of thermoelectric materials, self-healing coatings, and magnetorheological fluids further supports the evolution of smart automotive designs. These materials help optimize energy consumption, minimize environmental impact, and extend vehicle lifespan by reducing wear and tear. The integration of biomimetic materials also enhances adaptability and aerodynamics, drawing inspiration from nature to create more efficient and sustainable transportation solutions.

By combining these advanced materials with cutting-edge automotive technologies, manufacturers are redefining vehicle functionality and user experience.

Looking ahead, the automotive industry's reliance on smart materials is expected to grow as demand for sustainable, high-performance vehicles increases. Future advancements will likely focus on improving the cost-effectiveness, durability, and multifunctionality of these materials, making them even more viable for widespread adoption. As research continues to push the boundaries of material science, the next generation of automobiles will be smarter, lighter, and more energy-efficient, driving the industry toward a new era of intelligent transportation.

References

1. Chen Q, Kalpoe T, Jovanova J (2024) Design of mechanically intelligent structures: review of modelling stimuli-responsive materials for adaptive structures. Heliyon 10(14):e34026. https://doi.org/10.1016/j.heliyon.2024.e34026
2. Qader İN, Kök M, Dagdelen F, Aydoğdu Y (2019) A review of smart materials: researches and applications. El-Cezeri J Sci Eng 6(3):755–788. https://doi.org/10.31202/ecjse.562177
3. Global automotive smart materials market by type (interior, external), by application (commercial vehicle, private vehicle), by geographic scope and forecast. https://www.verifiedmarketreports.com/product/automotive-smart-materials-market-size-and-forecast/. Accessed 28 Feb 2025
4. Ammar M, Haleem A, Javaid M, Bahl S, Garg SB, Shamoon A, Garg J (2022) Significant applications of smart materials and Internet of Things (IoT) in the automotive industry. Mater Today Proc 68(Part 5):1542–1549. https://doi.org/10.1016/j.matpr.2022.07.180
5. Shukla U, Garg K (2023) Journey of smart material from composite to shape memory alloy (SMA), characterization and their applications-a review. Smart Mater Med 4:227–242. https://doi.org/10.1016/j.smaim.2022.10.002
6. The materials that drive the automotive industry. https://www.goodfellow.com/usa/resources/the-materials-that-drive-the-automotive-industry/. Accessed 28 Feb 2025
7. How can you use smart materials to improve automotive design? https://www.linkedin.com/advice/0/how-can-you-use-smart-materials-improve-automotive. Accessed 28 Feb 2025
8. Chandra Sekhar B, Dhanalakshmi B, Srinivasa Rao B, Ramesh S, Venkata Prasad K, Subba Rao PSV, Parvatheeswara Rao B (2021) Piezoelectricity and its applications. IntechOpen, UK. https://doi.org/10.5772/intechopen.96154
9. Dayyoub T, Maksimkin AV, Filippova OV, Tcherdyntsev VV, Telyshev DV (2022) Shape memory polymers as smart materials: a review. Polymers 14:3511. https://doi.org/10.3390/polym14173511
10. Thermoelectric materials. https://www.mpie.de/4297523/thermoelectric_materials. Accessed 28 Feb 2025
11. Choi K, Noh A, Kim J, Hong PH, Ko MJ, Hong SW (2023) Properties and applications of self-healing polymeric materials: a review. Polymers 15:4408. https://doi.org/10.3390/polym15224408
12. Self-healing materials in the automotive industry of the future. https://www.azom.com/article.aspx?ArticleID=22989. Accessed 28 Feb 2025
13. Ikura R, Park J, Osaki M et al (2022) Design of self-healing and self-restoring materials utilizing reversible and movable crosslinks. NPG Asia Mater 14:10. https://doi.org/10.1038/s41427-021-00349-1
14. Self-healing materials systems: overview of major approaches and recent developed technologies. https://onlinelibrary.wiley.com/doi/10.1155/2012/854203. Accessed 28 Feb 2025

15. Practical applications of self-healing polymers beyond mechanical and electrical recovery. https://onlinelibrary.wiley.com/doi/full/10.1002/advs.202302463. Accessed 28 Feb 2025
16. Buschow KHJ, de Boer FR (2003) Magnetostrictive materials. In: Physics of magnetism and magnetic materials. Springer, Boston, MA. https://doi.org/10.1007/0-306-48408-0_16
17. Osial M, Pregowska A, Warczak M, Giersig M (2023) Magnetorheological fluids: a concise review of composition, physicochemical properties, and models. J Intell Mater Syst Struct 34(16):1864–1884. https://doi.org/10.1177/1045389X231157357
18. Cannavale A, Ayr U, Fiorito F, Martellotta F (2020) Smart electrochromic windows to enhance building energy efficiency and visual comfort. Energies 13:1449. https://doi.org/10.3390/en13061449
19. Ciulla MG, Massironi A, Sugni M, Ensign MA, Marzorati S, Forouharshad M (2023) Recent advances in the development of biomimetic materials. Gels 9:833. https://doi.org/10.3390/gels9100833
20. Zeng H, Wang Y, Jiang T, Xia H, Gua X, Chen H (2021) Recent progress of biomimetic motions—from microscopic micro/nanomotors to macroscopic actuators and soft robotics. RSC Adv 11:27406–27419. https://doi.org/10.1039/D1RA05021D
21. Shafique M, Luo X (2019) Nanotechnology in transportation vehicles: an overview of its applications, environmental, health and safety concerns. Materials 12:2493. https://doi.org/10.3390/ma12152493
22. Smith M, Kar-Narayan S (2021) Piezoelectric polymers: theory, challenges and opportunities. Int Mater Rev 67(1):65–88. https://doi.org/10.1080/09506608.2021.1915935
23. Supriya S (2022) A review on lead-free-$Bi_{0.5}Na_{0.5}TiO_3$ based ceramics and films: dielectric, piezoelectric, ferroelectric and energy storage performance. J Inorg Organomet Polym 32:3659–3676. https://doi.org/10.1007/s10904-022-02418-6
24. Piezoceramics in the automotive industry. https://www.ceramtec-industrial.com/en/products-applications/piezo-technology/products-applications/piezoceramics-in-the-automotive-industry. Accessed 28 Feb 2025
25. Kulkarni H, Zohaib K, Khusru A, Shravan Aiyappa K (2018) Application of piezoelectric technology in automotive systems. Mater Today Proc 5(10)Part 1:21299−21304. https://doi.org/10.1016/j.matpr.2018.06.532
26. Automotive piezoelectric applications. https://www.americanpiezo.com/markets/automotive.html. Accessed 28 Feb 2025
27. Application of piezoelectric sensors in automobiles. https://www.he-shuai.com/application-of-piezoelectric-sensors-in-automobiles/. Accessed 28 Feb 2025
28. Chen F (2023) Applications of piezoelectric materials in modern car industry and future trends. Bachelor's thesis, LUT School of Energy Systems
29. Battaglia M, Sellitto A, Giamundo A, Visone M, Riccio A (2023) Shape memory alloys applied to automotive adaptive aerodynamics. Materials 16:4832. https://doi.org/10.3390/ma16134832
30. Riccio A, Sellitto A, Ameduri S, Concilio A, Arena M (2021) Shape memory alloys (SMA) for automotive applications and challenges (Chap. 24). In: Concilio A, Antonucci V, Auricchio F, Lecce L, Sacco E (eds) Shape memory alloy engineering, Second edn. Butterworth-Heinemann, UK, pp 785−808. https://doi.org/10.1016/B978-0-12-819264-1.00024-8
31. Jani JM, Leary M, Subic A (2014) Shape memory alloys in automotive applications. Appl Mech Mater 663:248–253. https://doi.org/10.4028/www.scientific.net/amm.663.248
32. Ghoniem E, Elmasry A, Azoti W, Elmarakbi M, Fu Y, et al (2023) Shape memory alloys (SMAs) based composites for automotive crashworthiness applications. In: Proceedings of the 1st international conference on lightweight materials, Milan at Politecnico di Milano, Italy, 11−12 May 2023
33. Elmasry A, Ghoniem E, Azoti W, Elmarakbi M, Fu Y et al (2023) Advanced shape memory hybrid composites for enhancing crashworthiness. In: Proceedings of the FISITA world congress, Barcelona, Spain, 12−15 September 2023
34. Stachiv I, Alarcon E, Lamac M (2021) Shape memory alloys and polymers for MEMS/NEMS applications: review on recent findings and challenges in design, preparation, and characterization. Metals 11:415. https://doi.org/10.3390/met11030415

35. Ni-Ti shape memory alloys and their constituents (2024). https://www.samaterials.com/content/ni-ti-shape-memory-alloys-and-their-constituents.html
36. Farber E, Zhu J-N, Popovich A, Popovich V (2020) A review of NiTi shape memory alloy as a smart material produced by additive manufacturing. Mater Today Proc 30(Part 3):761–767. https://doi.org/10.1016/j.matpr.2020.01.563
37. Parvizi S, Hashemi SM, Moein S (2022) 19 – NiTi shape memory alloys: properties. In: Thomas S, Behera A, Nguyen TA (ed) Micro and nano technologies, nickel-titanium smart hybrid materials. Elsevier, pp 399–426. https://doi.org/10.1016/B978-0-323-91173-3.00021-3
38. Alaneme KK, Okotete EA (2016) Reconciling viability and cost-effective shape memory alloy options – a review of copper and iron based shape memory metallic systems. Eng Sci Technol Int J 19(3):1582–1592. https://doi.org/10.1016/j.jestch.2016.05.010
39. Ahmad S, Hashmi AW, Singh J et al (2024) Innovations in additive manufacturing of shape memory alloys: alloys, microstructures, treatments, applications. J Mater Res Technol 32:4136–4197. https://doi.org/10.1016/j.jmrt.2024.08.213
40. Cohades A, Michaud V (2018) Shape memory alloys in fibre-reinforced polymer composites. Adv Ind Eng Polym Res 1(1):66–81. https://doi.org/10.1016/j.aiepr.2018.07.001
41. Bhekisipho T, Harsimran K, Amit T, Lovi Raj G, Sudhanshu D, Rajesh S (2024) A review of shape memory alloy applications and future trends in the aerospace industry. https://doi.org/10.2139/ssrn.4813587
42. Jani JM, Leary M, Subic A, Gibson MA (2014) A review of shape memory alloy research, applications and opportunities. Mater Des 56:1078–1113. https://doi.org/10.1016/j.matdes.2013.11.084

MIX
Papier aus verantwortungsvollen Quellen
Paper from responsible sources
FSC® C105338

If you have any concerns about our products,
you can contact us on
ProductSafety@springernature.com

In case Publisher is established outside the EU,
the EU authorized representative is:
Springer Nature Customer Service Center GmbH
Europaplatz 3, 69115 Heidelberg, Germany

Printed by Libri Plureos GmbH
in Hamburg, Germany